# Electrophysiological Maneuvers for Arrhythmia Analysis

# Electrophysiological Maneuvers for Arrhythmia Analysis

Written and Edited By:

**George J. Klein**, MD, FRCP(C); Professor of Medicine,
Division of Cardiology, Western University, London, Ontario, Canada

Cardiotext Publishing, LLC
750 2nd St NE Suite 102
Hopkins, MN 55343
USA
www.cardiotextpublishing.com

Any updates to this book may be found at: www.cardiotextpublishing.com/electrophysiological-maneuvers

Comments, inquiries, and requests for bulk sales can be directed to the publisher at: info@cardiotextpublishing.com.

Library of Congress Control Number: 2014937499

ISBN: 978-1-935395-89-8          6 7 8 9 10
eISBN: 978-1-935395-15-7

# Table of Contents

# Contributors

## Written and Edited By:

**George J. Klein**, MD, FRCP(C); Professor of Medicine, Division of Cardiology, Western University, London, Ontario, Canada

## Contributors:

**Lorne J. Gula**, MD, FRCP(C); Associate Professor of Medicine, Division of Cardiology, Western University, London, Ontario, Canada

**Peter Leong-Sit**, MD, FRCP(C); Assistant Professor of Medicine, Division of Cardiology, Western University, London, Ontario, Canada

**Jaimie Manlucu**, MD, FRCP(C); Assistant Professor of Medicine, Division of Cardiology, Western University, London, Ontario, Canada

**Paul D. Purves**, BSc, RCVT, CEPS; Senior Electrophysiology Technologist, Cardiac Investigation Unit, London Health Sciences Centre, London, Ontario, Canada

**Allan C. Skanes**, MD, FRCP(C); Professor of Medicine, Division of Cardiology, Western University, London, Ontario, Canada

**Raymond Yee**, MD, FRCP(C); Professor of Medicine, Director of Arrhythmia Service, Division of Cardiology, Western University, London, Ontario, Canada

# Foreword

While the field of electrophysiology has grown as a discipline, much of the focus of current trainees has been on interventional electrophysiology. There has been an erosion of education related to the underpinnings of electrophysiology: pathophysiology and mechanisms of arrhythmias. *Electrophysiological Maneuvers for Arrhythmia Analysis* is a unique attempt to organize, in a practical way, methodologies that can be used to understand tachycardia mechanisms and origins. There is no other text that aims to achieve these goals. Dr. Klein is one of the senior clinical electrophysiologists in the world and a world-class educator. This book imparts Dr. Klein's vast experience in the field in which he has made seminal contributions, as well as his expertise as an educator.

One of the attributes of Dr. Klein's approach to teaching electrophysiologic principles is that he relates the physiology to specific clinical problems. He has chosen a number of coauthors, with whom he has shared years of experience, to help provide a concise, practical, and easily understandable guide to allow physicians to approach electrophysiologic problems effectively. Despite the fact that many of these concepts are difficult, the authors have been able to make the explanations of the concepts clear and simple. They explain how stimulation, pharmacologic and physiologic perturbations can resolve important questions related to the differential diagnosis of tachycardia mechanisms. Their approach to analyzing arrhythmias is systematic and straightforward. Clinical examples are given for all the "maneuvers." The benefits and limitations of each are addressed in great detail from the opening chapter, in which Dr. Klein provides an overview of the book. The remaining chapters discuss the methods used to define sites of origin of arrhythmia, the presence or absence of preexcitation, the role of bypass tracts with an AV node in arrhythmias, and how to determine where unexpected signals arise (e.g., pulmonary vein vs. left atrial appendage).

In my opinion, this book should be on the shelf of every electrophysiologist trainee as well as every clinical cardiac electrophysiologist. It is a classic, like its editor. Dr. Klein deserves high praise for organizing his and his colleagues' clinical experiences and thought processes into a concise, practical text that should be part of all training programs in electrophysiology.

—*Mark E. Josephson, MD*

*Chief, Division of Cardiovascular Medicine; Director, Harvard-Thorndike Electrophysiology Institute and Arrhythmia Service, Beth Israel Deaconess Medical Center; Herman Dana Professor of Medicine, Harvard Medical School, Boston, Massachusetts*

# Preface

The last two decades have witnessed an explosion of technology to facilitate ablation of a wide variety of arrhythmias. Much of this has been in sophisticated imaging technologies and mapping of arrhythmias. Notwithstanding this, there remains an integral role for "traditional" electrogram (EGM) analysis for the simple reason that it is useful to understand the mechanism of a tachycardia to plan an intelligent ablation strategy.

The key to understanding the mechanism of a complex tachycardia often lies in watching its behavior in the face of perturbations, and this is the fundamental rationale underpinning the "electrophysiogical maneuver." To this end, there have been many detailed and lucid publications describing these, and the purpose of this publication is not to create an encyclopedic volume of them. Our laboratory has been training electrophysiology fellows for more than 30 years, and we have watched many highly gifted (and some not so!) trainees pass through. It is our view that the key to understanding electrophysiologic data lies in an understanding of a few relatively simple unifying concepts as we detail in chapter 1. The task is then to perform the maneuver properly and to apply a systematic approach to the understanding of what is observed. Understanding the data should not depend on an intuitive ability to visualize all complexities at a glance. We present typical examples and scenarios to illustrate this systematic approach to make the diagnosis and hopefully avoid the diagnostic pitfalls.

# Abbreviations

| | |
|---|---|
| **AF** | atrial fibrillation |
| **AP** | accessory pathway |
| **AT** | atrial tachycardia |
| **AV** | atrioventricular |
| **AVN** | AV node |
| **AVNRT** | AV nodal reentrant tachycardia |
| **AVRT** | AV reentrant tachycardia |
| | |
| **CL** | cycle length |
| **CS** | coronary sinus |
| | |
| **EGM** | electrogram |
| **EP** | electrophysiology |
| **ERP** | effective refractory period |
| | |
| **HB** | His bundle |
| **HRA** | high right atrium; high right atrial electrogram |

| | | | | |
|---|---|---|---|---|
| **JT** | junctional tachycardia | | **S, St, Stim** | stimulus |
| | | | **SA** | sinoatrial |
| **LBB** | left bundle branch | | **SVT** | supraventricular tachycardia |
| LBBB | left bundle branch block | | | |
| **LIPV** | left inferior pulmonary vein | | **TCL** | tachycardia cycle length |
| **LSPV** | left superior pulmonary vein | | **TVA** | tricuspid valve annulus |
| **LV** | left ventricle | | | |
| | | | **VA** | ventriculoatrial |
| **ms** | millisecond | | **VF** | ventricular fibrillation |
| | | | **VT** | ventricular tachycardia |
| **PCL** | pacing cycle length | | | |
| **PPI** | postpacing interval | | **WPW** | Wolff-Parkinson-White |
| **PVC** | premature ventricular contraction | | | |
| **PV** | pulmonary vein | | | |
| | | | | |
| **RBB** | right bundle branch | | | |
| **RBBB** | right bundle branch block | | | |
| **RIPV** | right inferior pulmonary vein | | | |
| **RSPV** | right superior pulmonary vein | | | |
| **RV** | right ventricle | | | |
| **RVA** | right ventricular apex | | | |

# Principles of the Electrophysiological Maneuver

Observations during ongoing tachycardia or during induction or termination of the arrhythmia usually expose the mechanism of a given arrhythmia during electrophysiologic study. Nonetheless, there may not be a diagnostic observation that definitively distinguishes all possibilities. In such instances, it is necessary to invoke one or more key interventions that might be expected to arbitrate among the diagnostic possibilities for a given arrhythmia.

This might be a stimulation maneuver (usually), pharmacological maneuver, or other. The maneuver might be during ongoing tachycardia or during sinus rhythm or pacing. The maneuver during tachycardia will be most definitive but maneuvers during sinus or paced rhythms may establish the physiological background that enhances or diminishes the probability of a given mechanism. For example, the para-His bundle pacing maneuver can establish the presence or absence of an accessory pathway (AP) and narrow the diagnostic possibilities for a given arrhythmia, although it doesn't directly prove that the AP is involved in the tachycardia.

Regardless of the tachycardia (wide QRS, atrial, supraventricular, etc.), the "drill" is always the same. That is, identify the universe of possibilities for a given dilemma and test the hypotheses with one or more maneuvers as necessary. There are clean "smoking gun" maneuvers but any maneuver is fallible for multiple reasons (operator error, ambiguous electrograms (EGMs), in diseased tissue, poor catheter access to a potentially key zone, capricious coincidence) and it is reassuring to verify by multiple means or to have a "tool box" of maneuvers.

# The Importance of "Geography"

It might seem obvious but factors like distance, conduction time, conduction barriers (both relative and absolute), and substrate all need to be considered and have a critical bearing in the thought process. Two EGMs can be far apart temporally because they are physically far apart, because there is conduction slowing between them or possibly because there is a conduction barrier between them and the route from one to another involves a long detour. Simple pacing maneuvers can distinguish these, as will be illustrated subsequently.

A simple application of "geography" can be illustrated in the following example. Our problem involves a patient with persistent retrograde conduction after ablation of a manifest posteroseptal pathway. Tachycardia is no longer inducible but retrograde conduction persists and is not cycle length dependent. Is the pathway still there? What simple maneuver can be done to test this relatively common dilemma? The ventriculoatrial (VA) conduction time, given a constant cycle length and constant atrial sampling site, will vary at different ventricular pacing sites, being longer the farther the pacing site is from the "exit" of the ventricles to the atrium. Retrograde conduction over the normal VA conduction system enters via the distal bundle branches, that is, more apically. With a functioning AP of the usual type, VA conduction will be over the AP, and the VA time will be shortest from the ventricular pacing site closest to the atrioventricular (AV) ring at the location of the AP.

This is illustrated in **Figure 1.1**, which shows that VA conduction time, when pacing closer to the base of the heart nearer the expected AP site P2, provides a shorter VA time (right panel) than pacing from the apical region P1 (left panel). This is most compatible with the continuing presence of an AP, which was the case. One would expect the opposite, if retrograde conduction were proceeding over the normal AV conduction system of which entrance is more apical (i.e., the distal bundle branches). Of course, there will be other ways to verify this, as will be evident in subsequent chapters.

It is timely here to consider the possible pitfalls of this specific maneuver, since similar ones apply for virtually any pacing maneuver. Consider what these might be. Remember that proof of the existence of the AP doesn't mean that it is involved in any given tachycardia. Consider the possibility that the pathway is cycle length dependent and has a long conduction time, causing it to be "later" than conduction over the normal system. Consider that it might have a long refractory period such that it is not conducting at the cycle length chosen for pacing. Consider that the AV node may have a very short conduction time over the retrograde system and compete with the pathway to get to the atria such that the difference in pacing site VA is attenuated. Consider finally that the pacing catheters may also not be exactly where you think they are. This partial list emphasizes the need to exercise care and judgment in interpreting any maneuver.

Figure 1.1

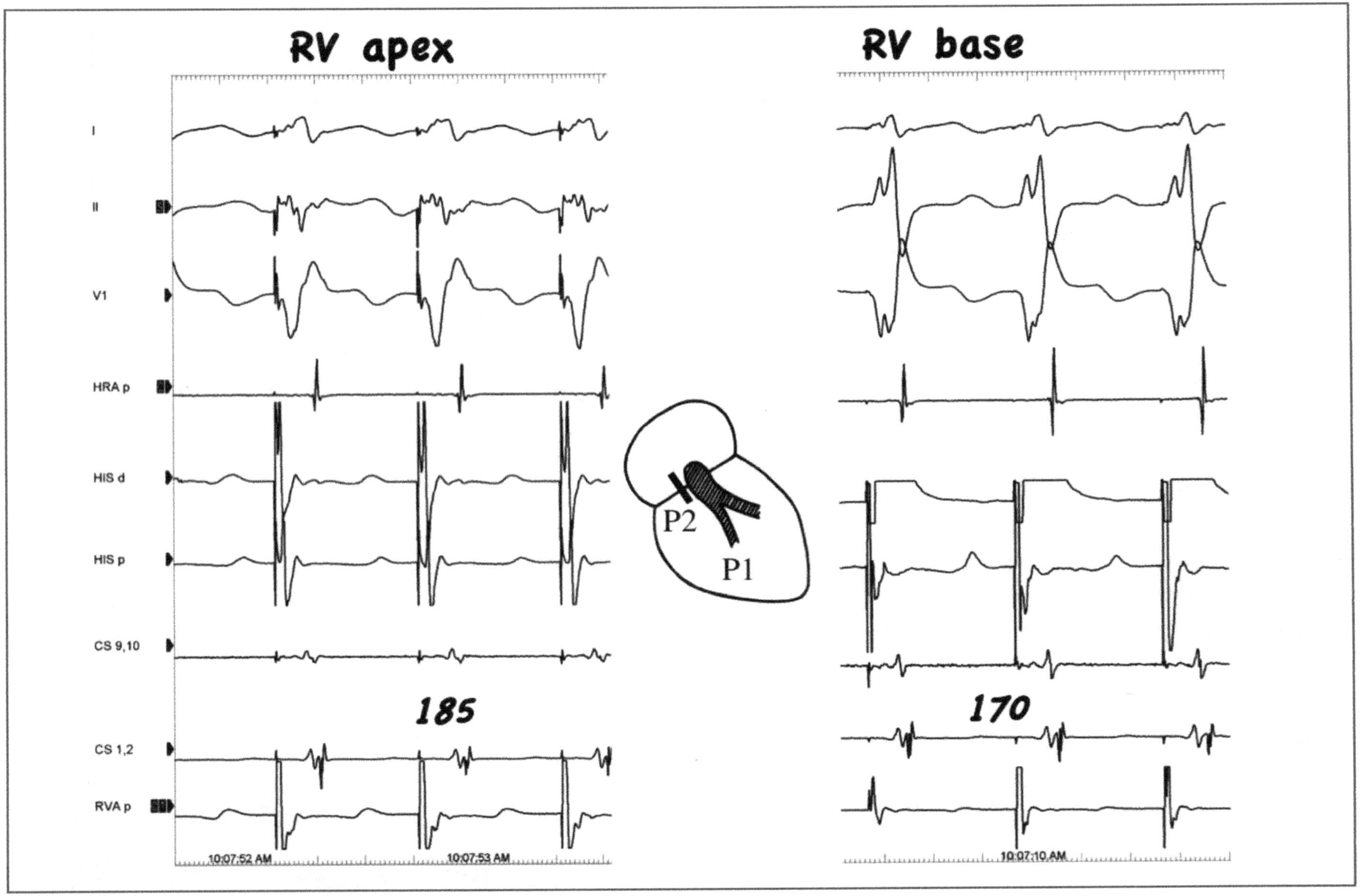
RV apex
RV base
I
II
V1
HRA p
HIS d
HIS p
CS 9,10
CS 1,2
RVA p
185
170
P2
P1
10:07:52 AM
10:07:53 AM
10:07:10 AM

# Fusion and Reset: A Key Concept

There are a plethora of EP maneuvers and "new" maneuvers continue to be published. This is much less intimidating to the novice if one considers that the great majority are derivatives of a few simple principles such as that of "geography" described previously and the concept of "fusion and reset." One might note that arguably the most compelling problem for the electrophysiologist is ascertaining the arrhythmia mechanism. This doesn't mean necessarily at a molecular or even cellular level but, for practical purposes, the electrophysiologist needs to know the mechanism well enough to define critical tissue to be ablated while limiting collateral damage to normal tissues.

Most clinical tachycardias can be thought of as being either "reentrant" or "focal" (although clearly fibrillation, torsades de pointes, and others need to be thought of differently). "Focal" is not a "mechanism" but is a descriptor for a relatively small "point source" mechanism that may be automatic, triggered, or "micro" reentrant (of secondary consideration for the purpose of ultimate ablation). This is illustrated graphically in **Figure 1.2**, which highlights the gross dissimilarity of these 2 classes of mechanism and should suggest pacing interventions to differentiate the 2 without significant difficulty.

Figure 1.2A depicts a classical reentrant circuit (in 2 dimensions here for clarity) with anatomic boundaries and an excitable gap of nonrefractory tissue between the advancing "head" and receding refractory "tail." An extrastimulus advances depolarization toward the gap, resulting in "fusion" with depolarization from the tachycardia, which may (if there is sufficient depolarization penetration by the extrastimulus) be evident on the surface ECG or intracardiac EGMs. The depolarization resulting from the extrastimulus here is unable to penetrate the circuit to conduct *antidromically* over the circuit due to refractoriness. However, the same circuit may get into the gap *orthodromically* and preexcite or advance the "tail" of the circuit. This may advance, delay, or terminate the next cycle depending on the properties of the circuit. Such a circuit is said to be "reset," although it is possible that reset is not discernable if slowing in the circuit due to prematurity balances the precocity of the extrastimulus. **One can appreciate that coincidental reset and fusion are only possible if the circuit or underlying mechanism has an excitable gap with a "separate entrance and exit."** With very rare exceptions, "fusion and reset" is only seen with "macroreentry." One can also readily predict the determinants of reset. These include size of the excitable gap (as determined by conduction time and refractoriness), proximity of the pacing site physically to the excitable gap, and physical barriers (e.g., scar) between the pacing site and the excitable gap.

**Figure 1.2**

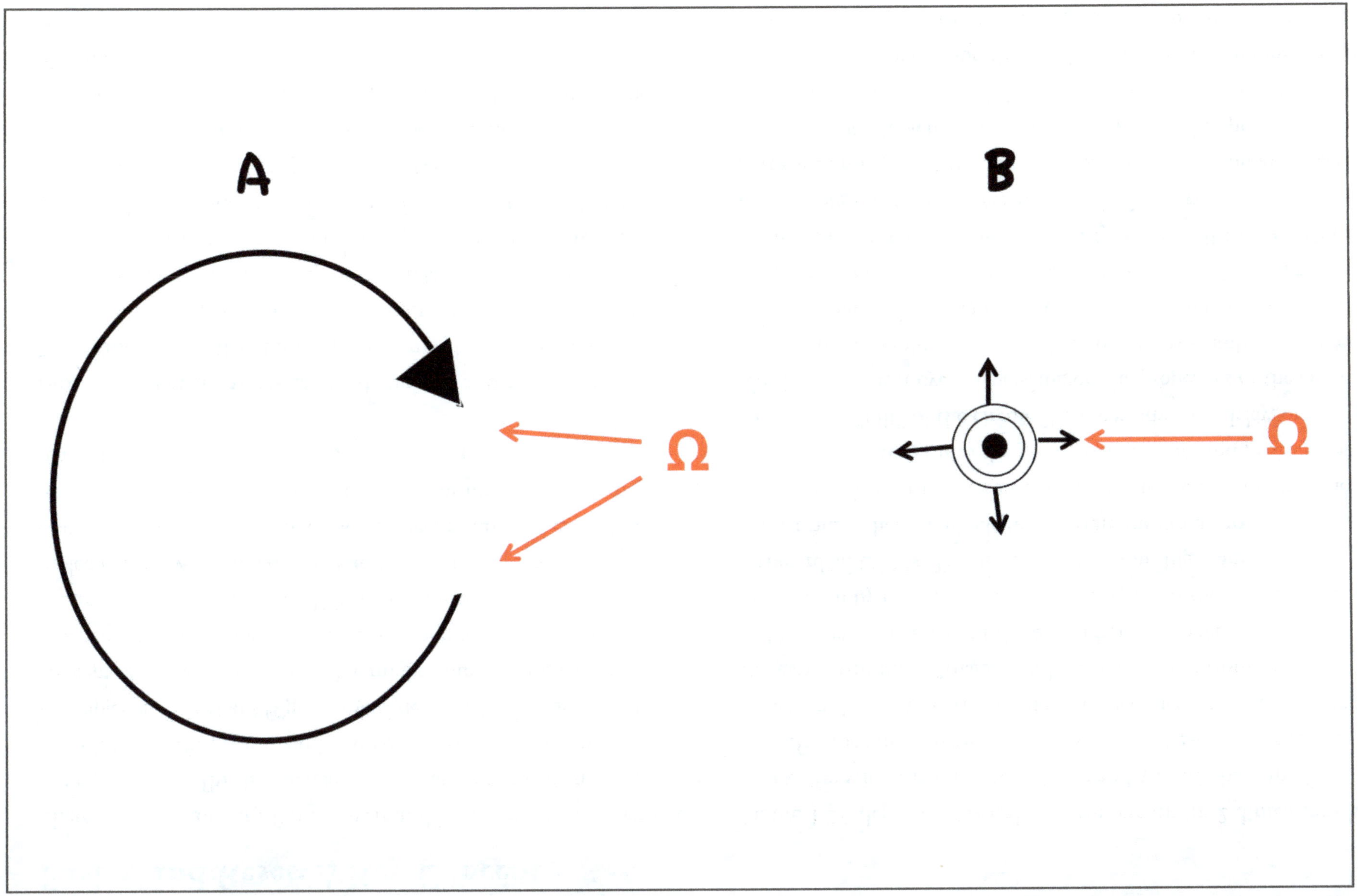

Consider the same extrastimulus in the case of the focal source of tachycardia in Figure 1.2B. It is easy to envisage fusion with collision of the wave from the pacing site with the wave from the focus. However, the wave from the focus in essence creates a protective barrier of refractoriness after discharge such that the tachycardia "generator" is not susceptible to disruption (i.e., reset). That is, *both* fusion *and* reset are not possible with a focus. Conversely, one can potentially penetrate the focus (reset) but it is only when there is no ring of refractoriness around the focus from its own depolarization, i.e., one can get reset without fusion but not both at the same time.

The importance of this concept is well illustrated with a clinical example using the quintessential model of macrorentry, namely atrioventricular reentrant tachycardia (AVRT) using an accessory AV pathway as part of the circuit.

This tracing (**Figure 1.3**) demonstrates a regular, narrow QRS tachycardia with a 1:1 AV relationship. The p is negative in lead II, suggesting low to high atrial activation. Earliest atrial activation is at the proximal coronary sinus but is relatively early at the His bundle and high right atrium (HRA) EGMs. A premature ventricular contraction (PVC) from the right ventricular apex (RVA) is delivered at a time when the His bundle activation is completed, i.e., the His deflection is on time. This *delays* the next atrial cycle, a situation that is only compatible with an accessory AV pathway as the retrograde limb of an AV reentrant circuit. This is the well-known "His refractory PVC" that is a fundamental of clinical electrophysiology predicated on the fact that the subsequent atrial depolarization can't be influenced via the normal AV conduction system if the His bundle is refractory. The "reset" indicates that the extrastimulus has penetrated the circuit to alter the subsequent cycle. In the preceding example, the extrastimulus has delayed rather than advanced the next cycle since the AP in question exhibits cycle length dependent prolongation of conduction time. This is illustrated in **Figure 1.4**. Most APs have a relative constant conduction time independent of cycle length (curve C), whereas some prolong the conduction as a function of more premature coupling of the extrastimulus (curve B). Whether the reset advances (#3 for both curves), delays (#2 for curve B) or terminates (#1 for both curves), the tachycardia depends on the degree of prematurity of the extrastimulus relative to the advancing wave front to the AP.

7

**Figure 1.3**

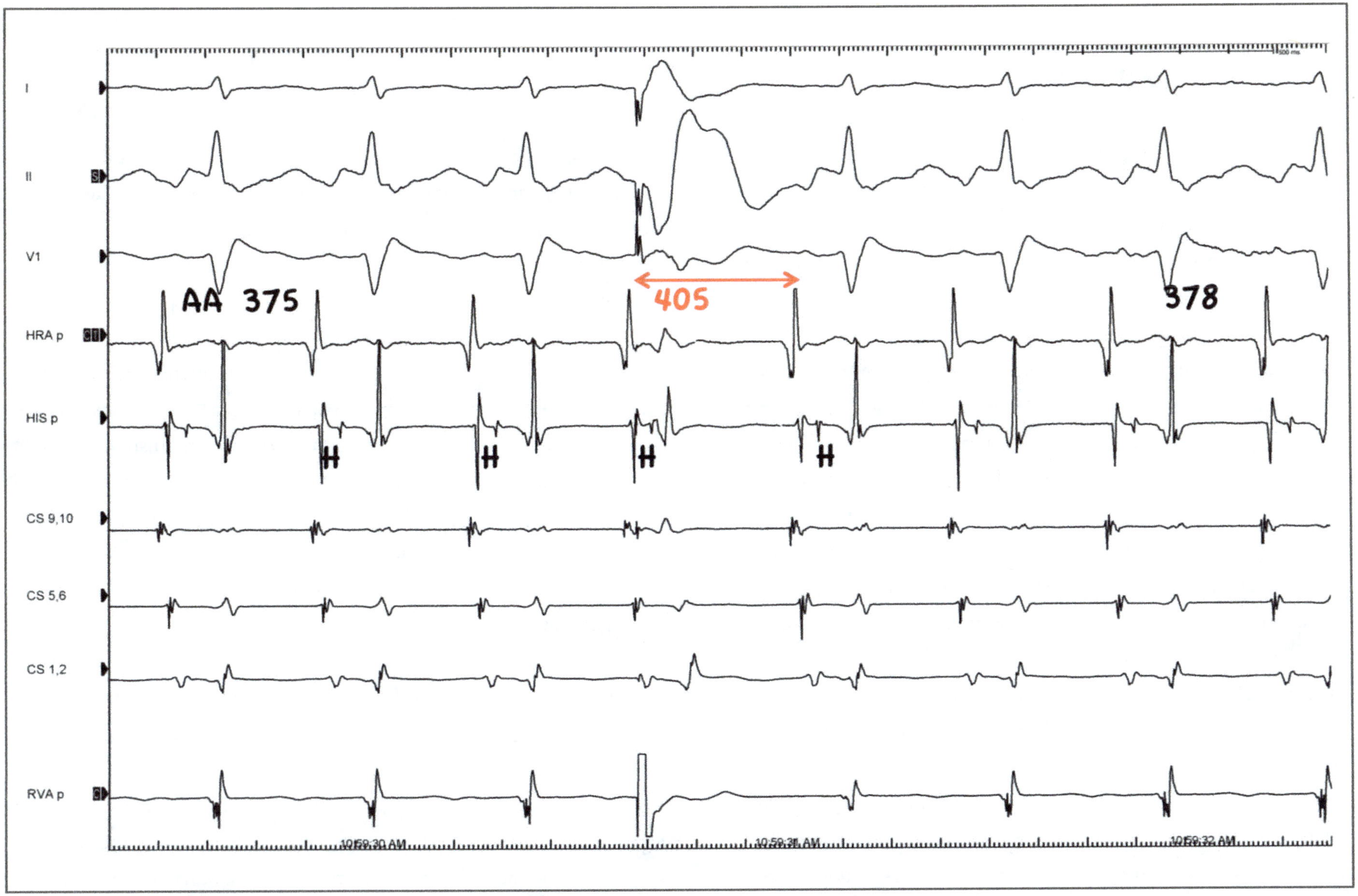

Figure 1.4

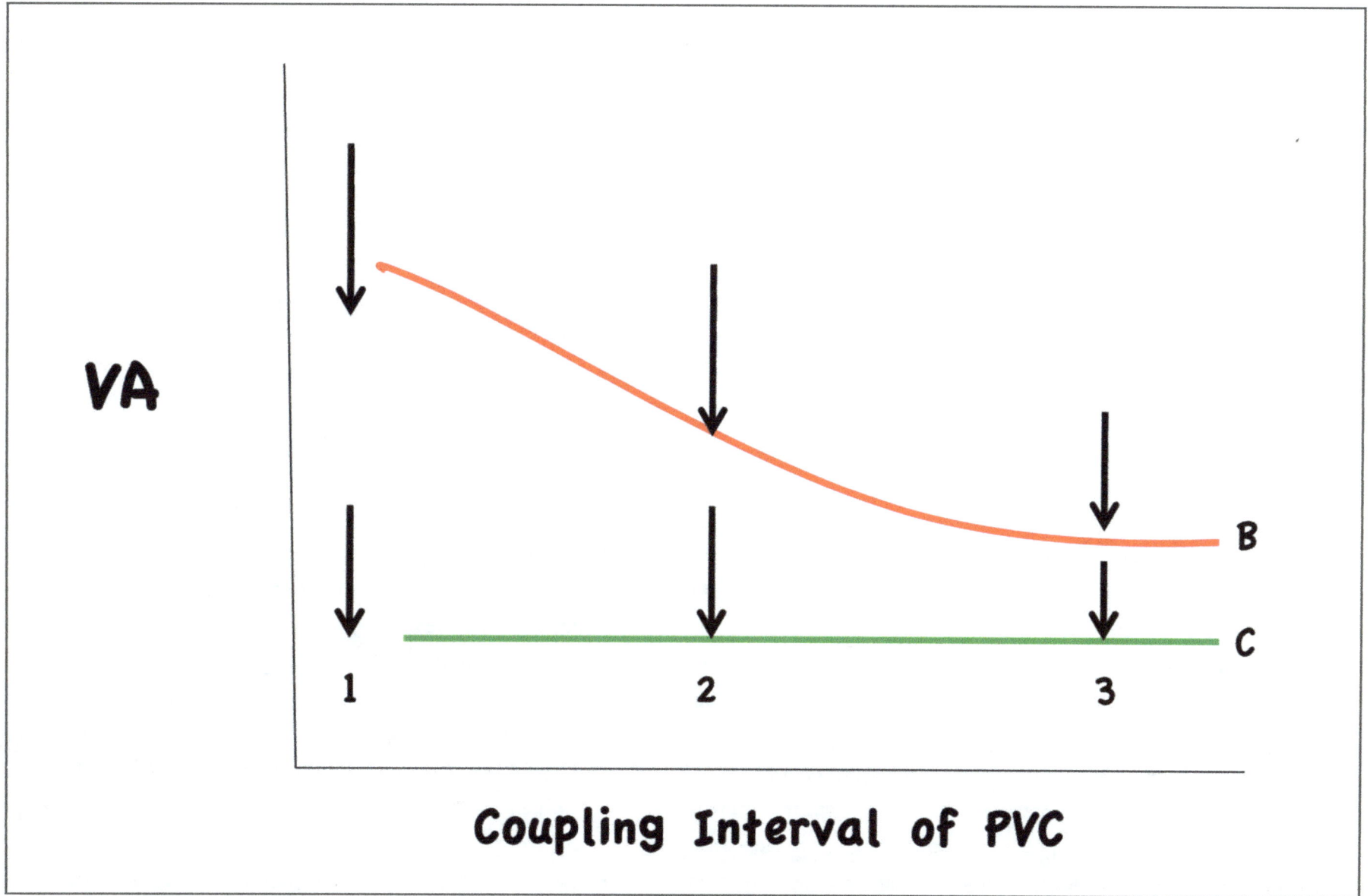

*It should become apparent that the "His refractory PVC" during AVRT is just a specific example of "fusion and reset."* Consider **Figure 1.5**, which only shows the surface ECG from the example in Figure 1.3. Atrial activity is evident (long VA interval) with a one-to-one AV relationship. The PVC is relatively late-coupled and delays both the subsequent P and QRS, that is, resets the tachycardia. In fact, the coupling interval of the PVC (305 ms) only preempts the next anticipated QRS by 70 ms, which, by a simple calculation (if the tachycardia is sufficiently regular), places the arrival of the anticipated anterograde wave over the normal AV conducting system 70 ms after the stimulus artifact. This must then be fused even if the "pure" paced QRS is not available for inspection (not shown, but the entirely paced QRS was indeed wider). The His bundle must be refractory on the arrival of the retrograde depolarization from the PVC as verified in the intracardiac record (Figure 1.3).

The "His refractory PVC" as an archetypal example of "fusion and reset" is pictorially illustrated in **Figure 1.6**, where the anterograde wave to the AP from the pacing stimulus (in red) reaches the AV node to reset the tachycardia, while concurrently the depolarization also advances to the bundle branches in the antidromic direction (in green) to the circuit to provide the fusion.

## Figure 1.5

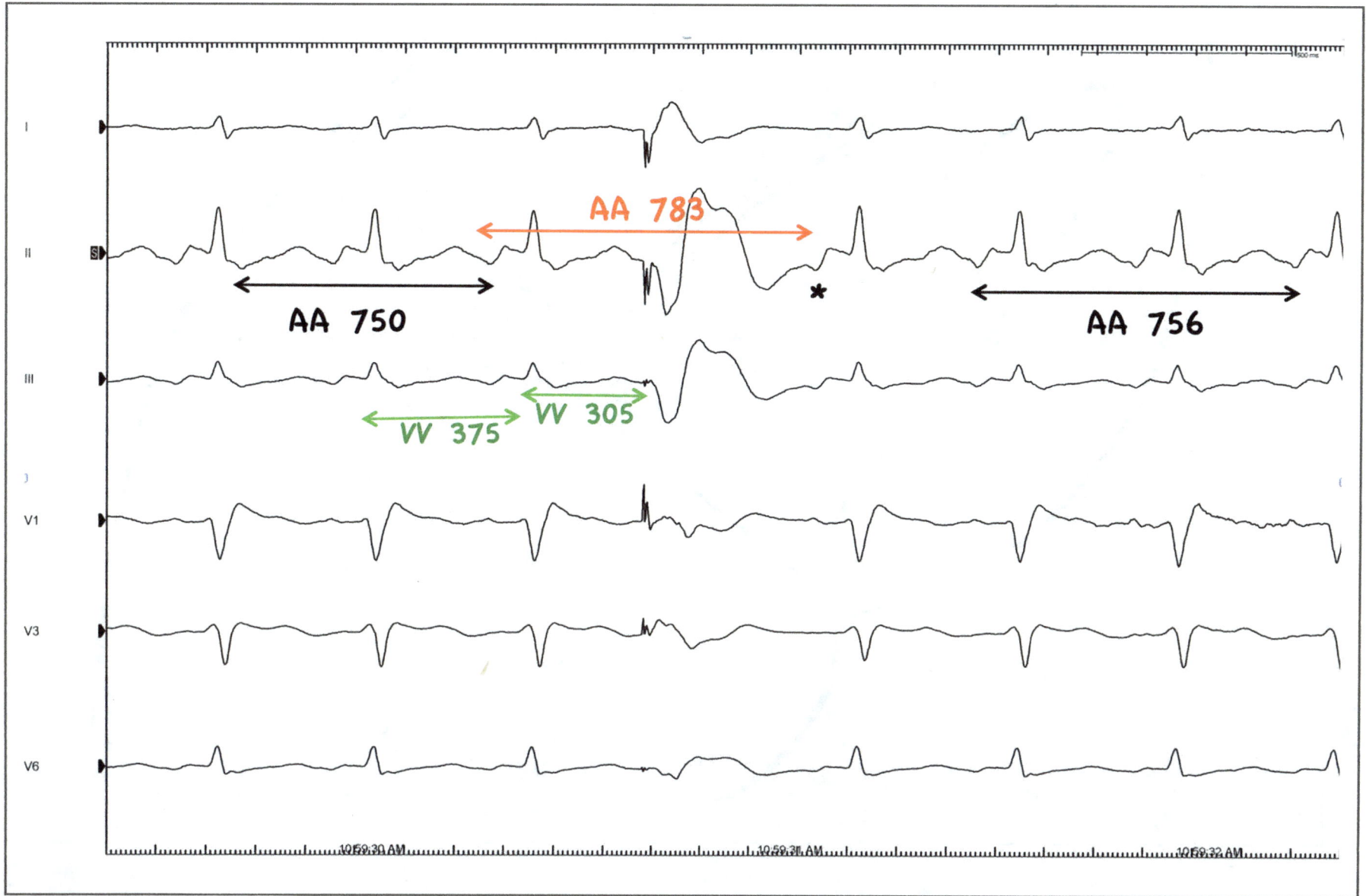

Figure 1.6

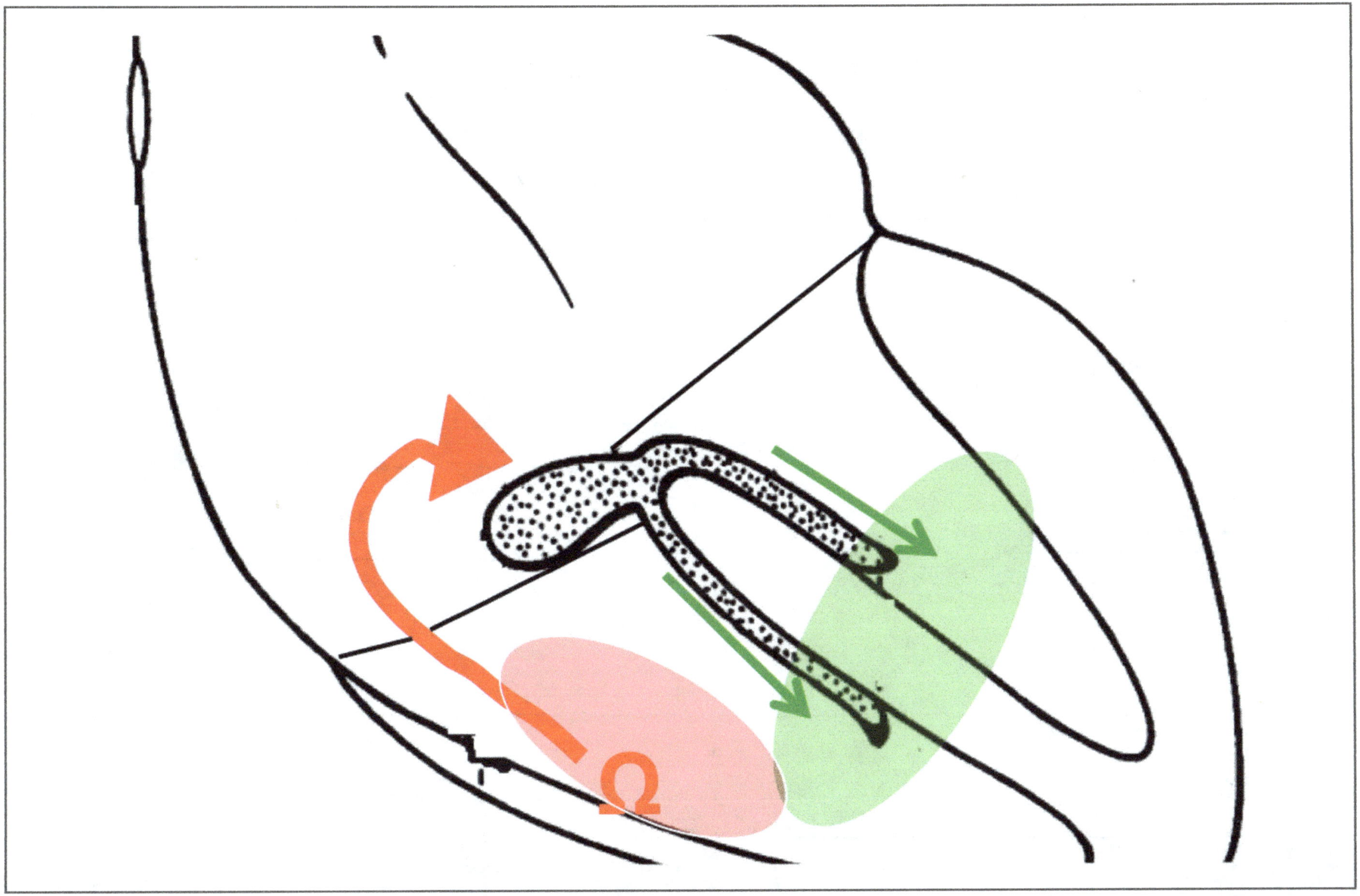

# The Relationship between the His Refractory PVC during Supraventricular Tachycardia (SVT) and Entrainment of SVT from the Ventricle

Overdrive pacing is universally done to assess mechanism of a regular SVT. Pacing from the ventricle is performed initially at a cycle length slightly shorter than that of tachycardia to minimize the potential for decremental conduction, which in a hypothetical circuit could confound conclusions. *Entrainment* is signaled by the observation of fixed QRS fusion at a given cycle length and verified by noting that this fixed fusion varies at different cycle lengths (Waldo criteria). (Constant fusion cannot be demonstrated with overdrive pacing of a focal tachycardia, which can only show the fully paced QRS morphology when the tachycardia has been overdriven. This will be discussed in more detail subsequently.) Suffice it to know here that entrainment from the ventricle requires *continuous* reset and fusion (as per Figure 1.2) and is really the "PVC during SVT" delivered repetitively. To repeat, *entrainment from the ventricle and delivery of a PVC into the tachycardia essentially constitute the same maneuver!* Conversely, the PVC delivered into SVT that advances is essentially *entrainment for one cycle*. Most of the time, as one would expect, they provide the same information. The PVC during tachycardia can scan the whole cardiac cycle and potentially provides a broader diagnostic swath than entrainment, since the latter is often only done at one cycle length of pacing with onset of pacing begun at one coupling interval to the ongoing tachycardia. Entrainment may potentially be more likely to terminate a tachycardia or be proarrhythmic. However, comparisons are moot since the experienced electrophysiologist will be familiar with both and pull them both out of the proverbial diagnostic bag in a difficult case.

The following will illustrate the close mechanistic and practical equivalence of 3 maneuvers. **Figure 1.7** shows these 3 "different" maneuvers in the same patient with SVT due to AV reentry over a right AV AP. One would expect the right ventricular (RV) pacing site to have excellent access to the excitable gap of the circuit for any of these. Figure 1.7A shows a PVC preexciting the subsequent atrial deflection when the His is refractory. The QRS is obviously fused and much narrower than one would expect with RV pacing alone even though a pure paced QRS is not available here for comparison. There is a "VAV" response, consistent with AV or AVN reentry. The postpacing interval after the PVC is only about 40 ms longer than the tachycardia cycle length and is therefore "in," consistent with AV reentry. The difference between the SA and the VA intervals is only 7 ms, again supporting AVRT. This is not surprising since the RV is "in" the circuit with AVRT over a right AP.

Figure 1.7B shows termination of pacing after entrainment. It is not surprising that the information is essentially identical to that derived from 1.7A, i.e., there is a VAV response, the postpacing interval (PPI) is "in," and the difference between the SA and VA is again minimal.

Finally, Figure 1.7C shows the onset of pacing during an entrainment attempt, a maneuver described initially by some as a novel maneuver. The first pacing stimulus that *captures* the ventricle is the third delivered, the next visibly shows QRS fusion ($F_1$) and the subsequent beat also shows fusion ($F_2$) but in addition advances ("resets") the A. This is clearly the same phenomenon as seen in Figures 1.7A and B. In AV reentry, it would be expected that "fusion and reset" be observed relatively early in the pacing sequence, if it is to occur, since the excitable gap is relatively big. By contrast, in AVN reentry, one would not see *both* fusion and reset (either one or the other), and reset would occur relatively later in the drive. Finally, one might note that the difference between the VA during tachycardia and the SA of the beat that advances (resets) the A is again very short, again as expected and essentially identical to that measured in Figures 1.7A and B.

**Figure 1.7A**

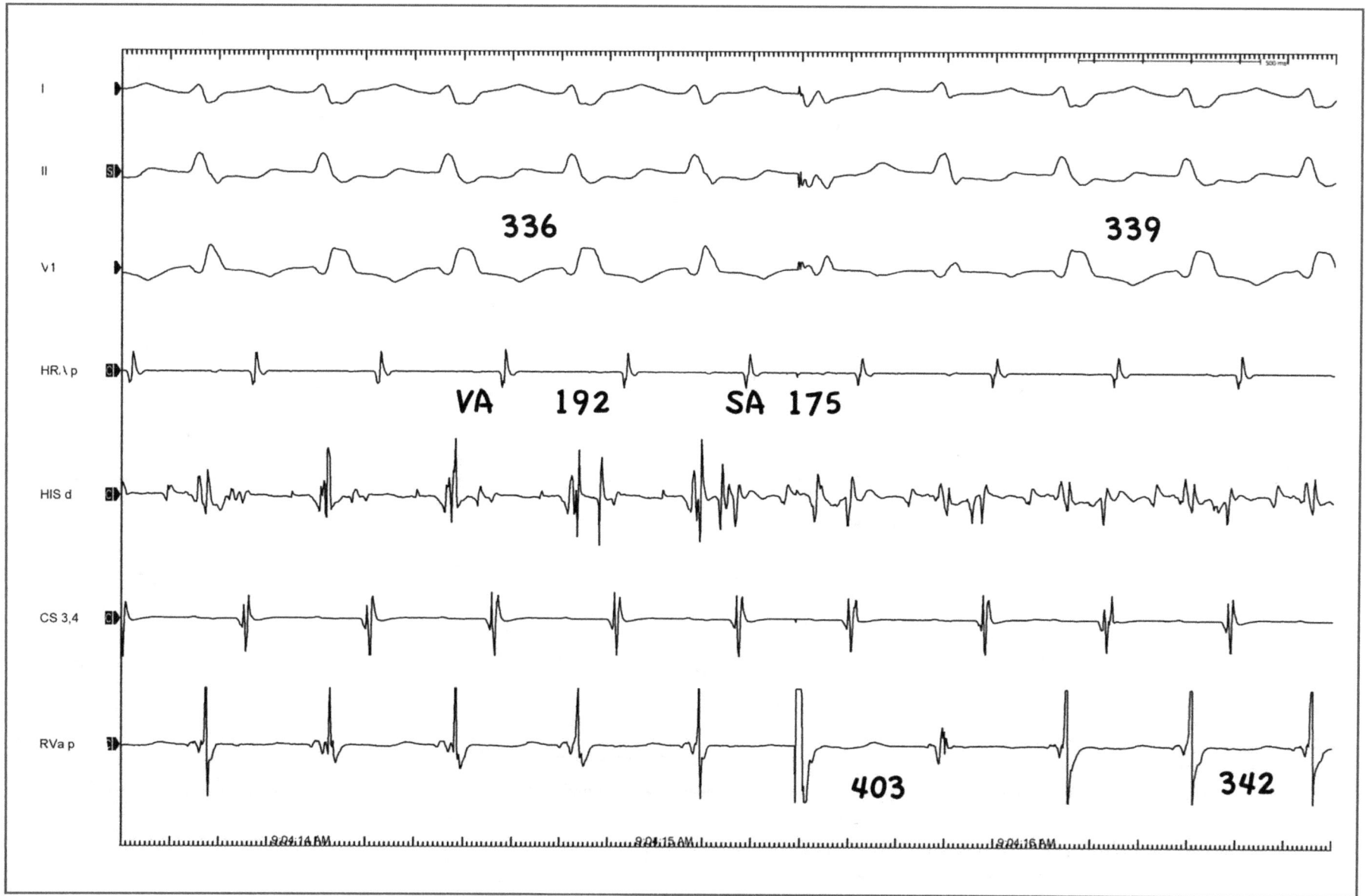

## Figure 1.7B

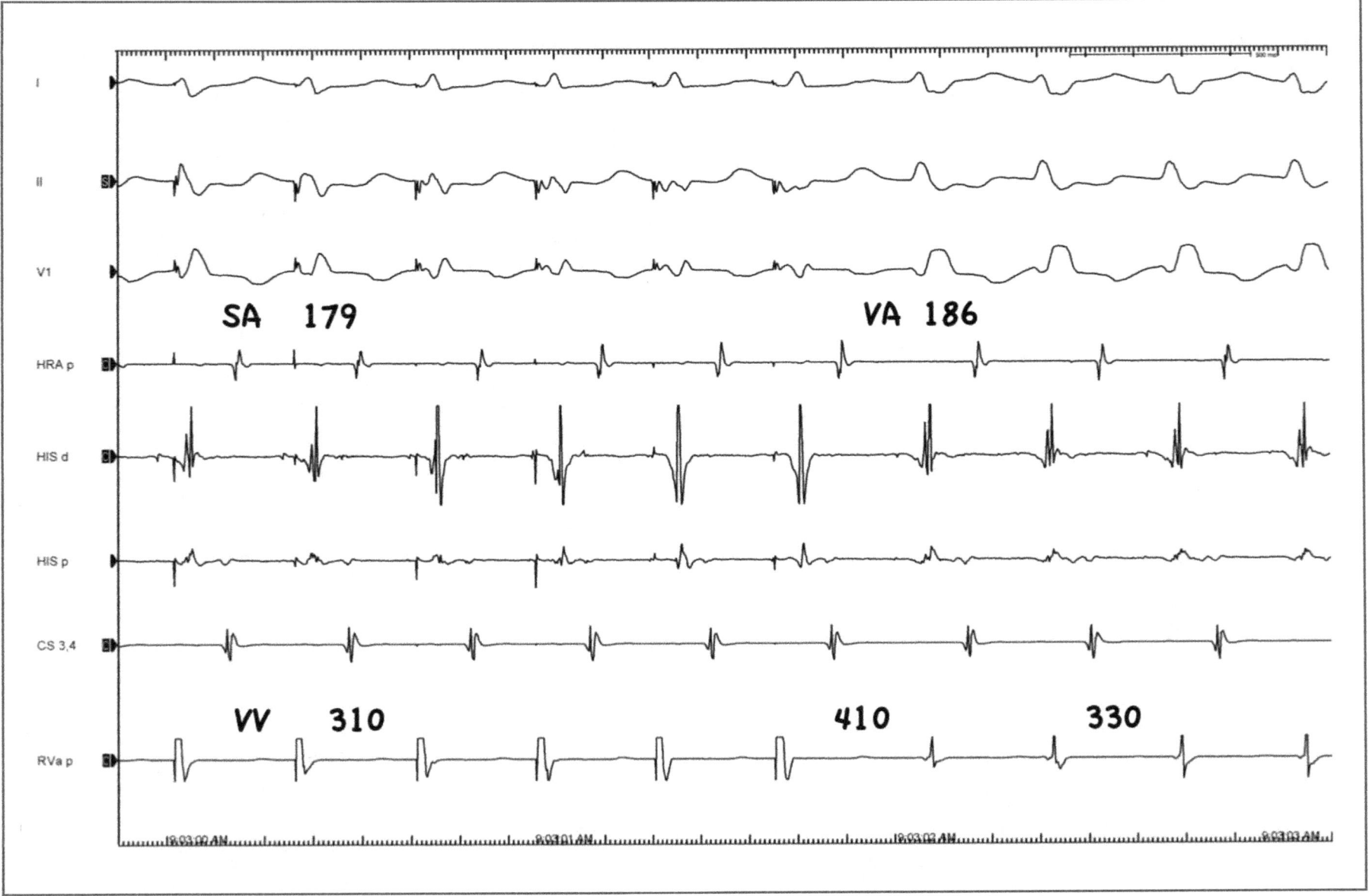

## Figure 1.7C

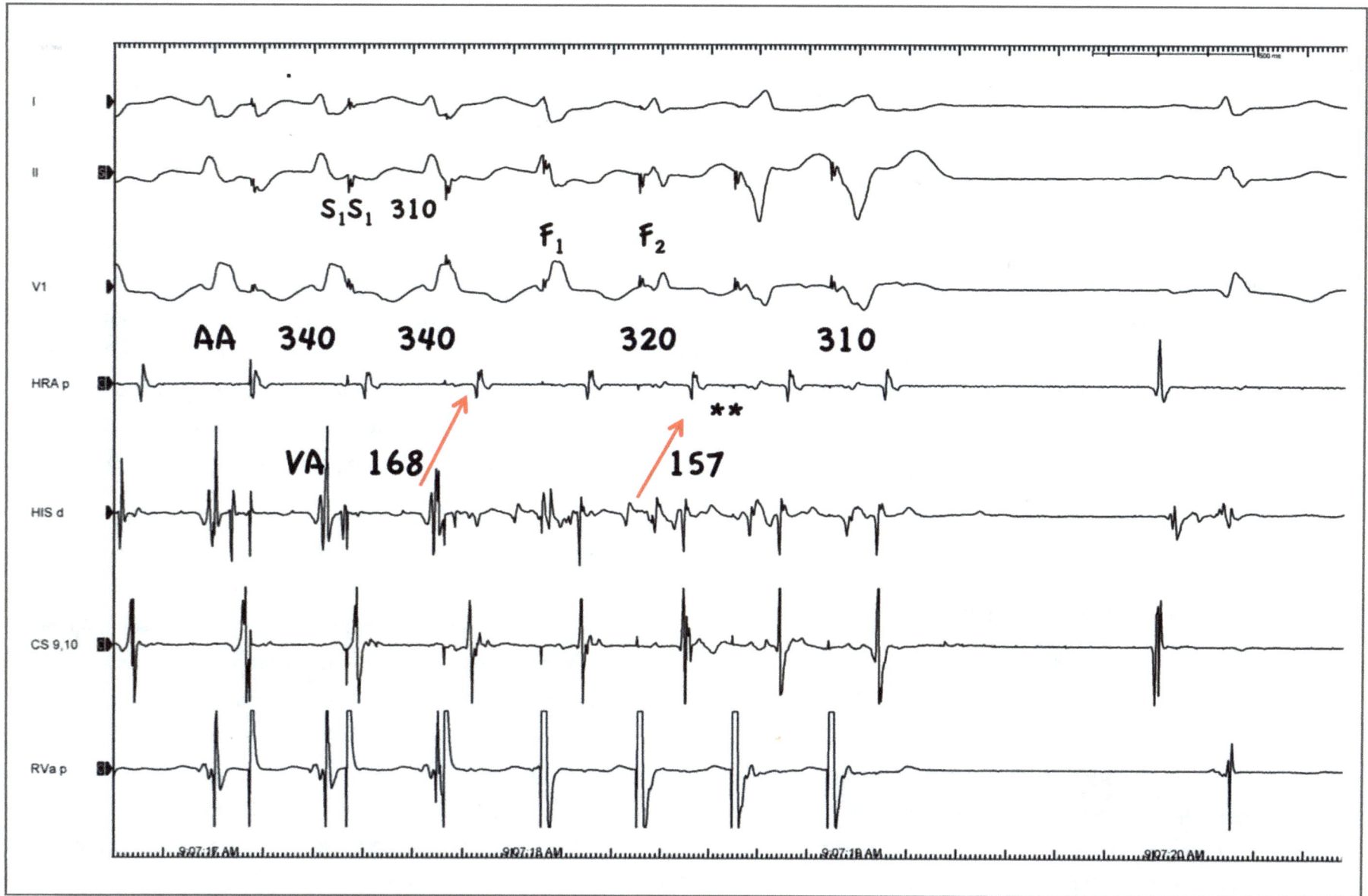

A few other generalities merit mention before a more detailed inspection of the maneuvers and specific examples that follow in subsequent chapters.

1. Pneumonics or "sound bites" such as VAV and VAAV abbreviating a maneuver are very handy in communication but must never be divorced from a fundamental understanding of the physiology behind a specific maneuver. There is no such thing as a simple rule and these *word traps* for the superficial observer are deflected by careful attention to what is happening electrophysiologically.

2. Certain maneuvers or rules based on them constitute virtually absolute proof of a mechanism or "smoking gun," while others are more probabilistic and need to be treated as such. For example, with a carefully done procedure, consistent preexcitation of the next A with a His refractory PVC can only occur over an AP. On the other hand, PPI greater than 115 ms or so after cessation of entrainment by RV apical pacing during SVT *suggests* AVN reentry, but there are notable exceptions as will be enumerated in subsequent chapters.

3. A "positive" response, such as advancement of A by a His refractory PVC, is proof of presence of an AP, but a negative response is not proof of absence of such. As another example,

PPI consistently equal to the cycle length of a tachycardia means that the pacing site is "in," as seen in Figure 1.7, but there are many other potential considerations when it is "out" by the numbers but actually "in" the circuit. One might summarize by saying that **"in" is "in" but "out" might also be "in."** This is a common mental mistake with novices.

4. Any maneuver is only as good as the care and attention to detail provided to perform it. It obviously needs to be reproducible, especially when there is some confounding cycle length variability. Anatomic anomalies, incomplete catheter coverage, and poor EGMs in diseased tissue will remain challenging in the best of hands.

My colleagues and I provide more detailed and specific examples of the various diagnostic interventions and their pitfalls in the subsequent chapters. The devil is often in the detail but the reader may wish to return to this introductory chapter at certain points as well as the end of the book to remind themselves of the "bigger picture."

# Suggested Readings

1. Wellens HJJ. Value and limitations of programmed electrical stimulation of the heart in the study and treatment of tachycardias. *Circulation*. 1978;57:845–853.

2. Waldo AL. From bedside to bench: entrainment and other stories. *Heart Rhythm*. 2004;1:94–106.

3. Waldo AL, MacLean WAH, Karp RB, Kouchoukas NT, James TN. Entrainment and interruption of atrial flutter with atrial pacing: studies in man following open heart surgery. *Circulation*. 1977;56:737–745.

4. Stevenson WG, Khan H, Sager P, et al. Identification of reentry circuit sites during catheter mapping and radiofrequency ablation of ventricular tachycardia late after myocardial infarction. *Circulation*. 1993;88:1647–1670.

5. Hirao K, Otomo K, Wang X, et al. Para-Hisian pacing: a new method for differentiating retrograde conduction over an accessory AV pathway from conduction over the AV node. *Circulation*. 1996;94:1027–1035.

6. Gallagher JJ, Gilbert M, Svenson R H, et al. Wolff-Parkinson-White syndrome. The problem, evaluation, and surgical correction. *Circulation*. 1975;51:767–785.

7. Bennett MT, Leong-Sit P, Gula LJ, et al. Entrainment for distinguishing atypical atrioventricular node reentrant tachycardia from atrioventricular reentrant tachycardia over septal accessory pathways with long right ventricular tachycardia. *Circ Arrhythm Electrophysiol*. 2011;4(4):506–509.

8. Knight BP, Ebinger M, Oral H, et al. Diagnostic value of tachycardia features and pacing maneuvers during paroxysmal supraventricular achycardia. *J Am Coll Cardiol*. 2000; 36:574–582.

# 2

# Maneuvers during Sinus or Paced Rhythm

Pacing maneuvers can be classified as perturbations during an ongoing tachycardia to shed light on the mechanism or those performed during sinus rhythm or pacing, usually to provide details of structure or function that provide insight into the potential tachycardias but would not directly diagnose the mechanism of an observed tachycardia. We discuss the latter in this chapter. A simple and straightforward example of this type of maneuver would be ventricular pacing with no retrograde conduction demonstrable at any cycle length or autonomic enhancement. This observation would make a junctional reentrant tachycardia very unlikely in the case of a clinical supraventricular tachycardia (SVT) and strongly support but not establish atrial tachycardia as the mechanism.

Many maneuvers are really intuitive if one visualizes the anatomy along with the tracings. **Figure 2.1A** is a 12-lead ECG clearly showing preexcitation with the pattern most suggestive of a posteroseptal accessory pathway (AP). As shown in **Figure 2.1B**, pacing from inside the coronary sinus (CS) changes the preexcitation pattern, and the tall R now seen in $V_1$ indicates conduction over a left AP as well as the posteroseptal one. This is a simple application of "geography" where pacing closer to the left pathway from the CS favored conduction over this pathway which was not apparent during sinus rhythm, whereas sinus node conduction favored conduction over the posteroseptal AP obscuring the left pathway. Pacing from more than one atrial site prior to a detailed electrophysiology (EP)

study for Wolff–Parkinson–White (WPW) syndrome is simple enough and can alert one earlier to the presence of multiple APs.

One can also refer to chapter 1, Figure 1.1 and the relevant discussion therein, where the ventriculoatrial (VA) interval was earlier with pacing at the base of the heart than that obtained from pacing at the apex (near the exit of the right bundle branch (RBB)). This is indicative of conduction over an AP, although it again does not prove the involvement of this pathway in any observed clinical tachycardia.

Finally, one might add that thinking anatomically (or "geographically") also applies to those maneuvers applied during tachycardia. For example, determining the mechanism of an atrial tachycardia often involves overdrive pacing from multiple atrial sites and it is not surprising that there is a direct relationship between the postpacing interval (PPI) and proximity to the circuit or source of tachycardia.

## Figure 2.1A

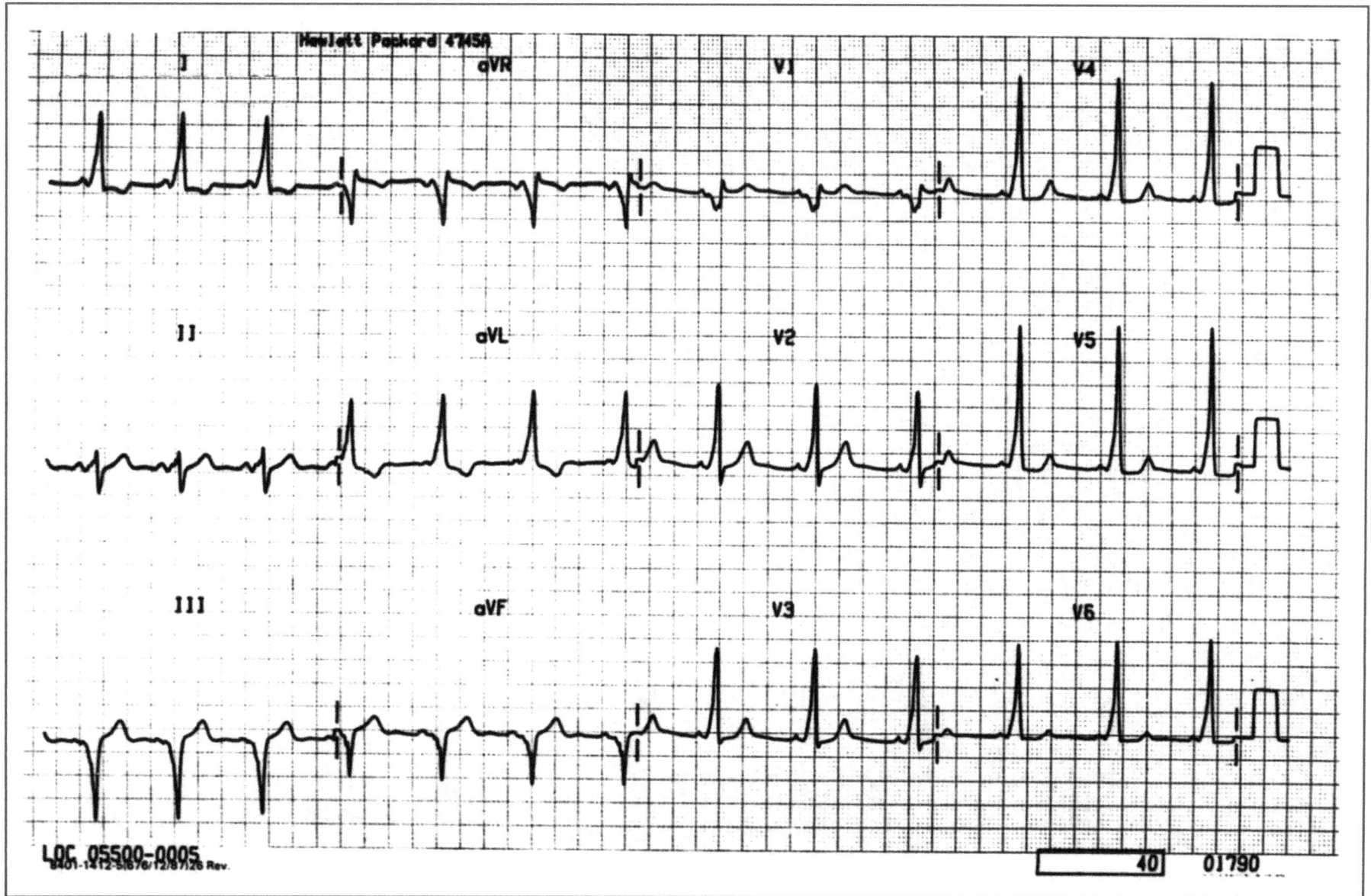

## Figure 2.1B

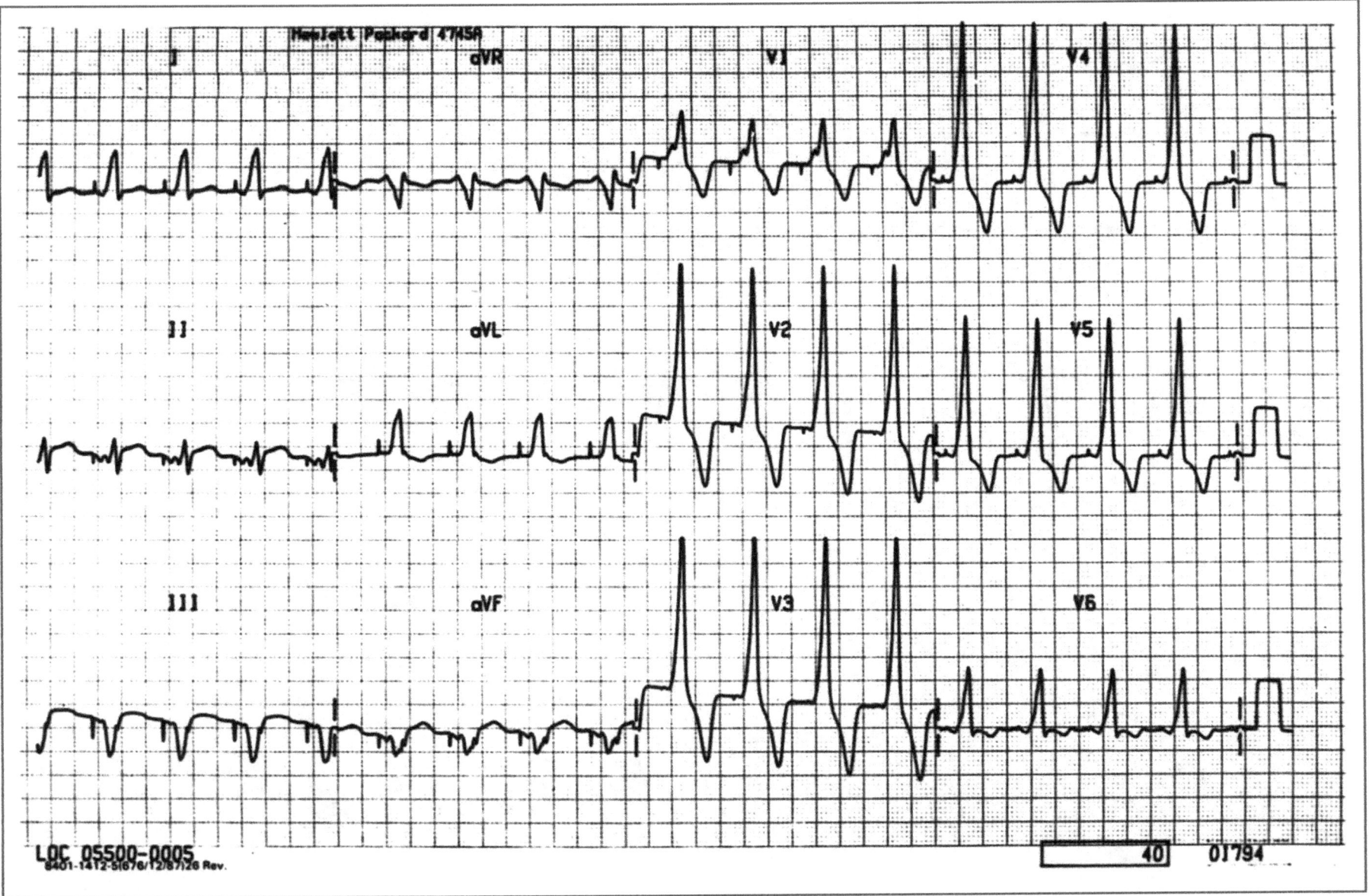

# Para-Hisian Pacing

Para-Hisian pacing is a useful maneuver, most frequently used to determine the presence of an AP in the septal and paraseptal region. It is generally performed in the context of SVT with a central retrograde atrial activation sequence where there is a question of atrioventricular (AV) node reentry vs. AV reentry utilizing a septal AP. Although the maneuver is conceptually easy to understand, interpretation of the maneuver and its many pitfalls sometimes prove to be more challenging.

The maneuver is performed by pacing from the bipolar electrodes positioned at the His position. The selected electrogram (EGM) components at the His site should record larger amplitude, near-field His and right ventricular (RV) EGMs and relatively smaller and more far-field atrial components to minimize inadvertent pacing of the atrium. High output (e.g., 20 V at 0.2 ms pulse width) is used to directly capture the bundle of His indicated by a relatively narrow QRS pacing morphology (identical to the supraventricular QRS, if ONLY the His bundle is captured). The output is then gradually lowered until a sudden increase in the width of the paced QRS morphology indicates loss of His capture with only para-Hisian ventricular capture remaining. It is optimal to record the His itself

from the proximal pair, but this is not necessary for the correct interpretation of the results.

It is always useful to return to or "visualize" **Figure 2.2** when interpreting a clinical example. Several things become apparent on observing this figure. They are:

1. There are three possible cardiac structures that can be directly captured by pacing in close proximity to the His region: (a) myocardium adjacent to the His bundle (para-Hisian); (b) the His bundle; and (c) the atrium adjacent to the His.

2. Para-Hisian and His capture refers to capture of both the His bundle and the adjacent ventricular myocardium (large square wave symbols on left panel of Figure 2.2).

3. His bundle capture usually results in anterograde and retrograde conduction via the normal AV conduction system resulting in relative early arrival of the impulse to the atrium and to the termini of the bundle branches (and hence the RV apex (RVA)). Thus, capture of the bundle of His is suggested by a narrower QRS morphology and, typically, a relatively short stimulation-to-RVA EGM interval.

Figure 2.2

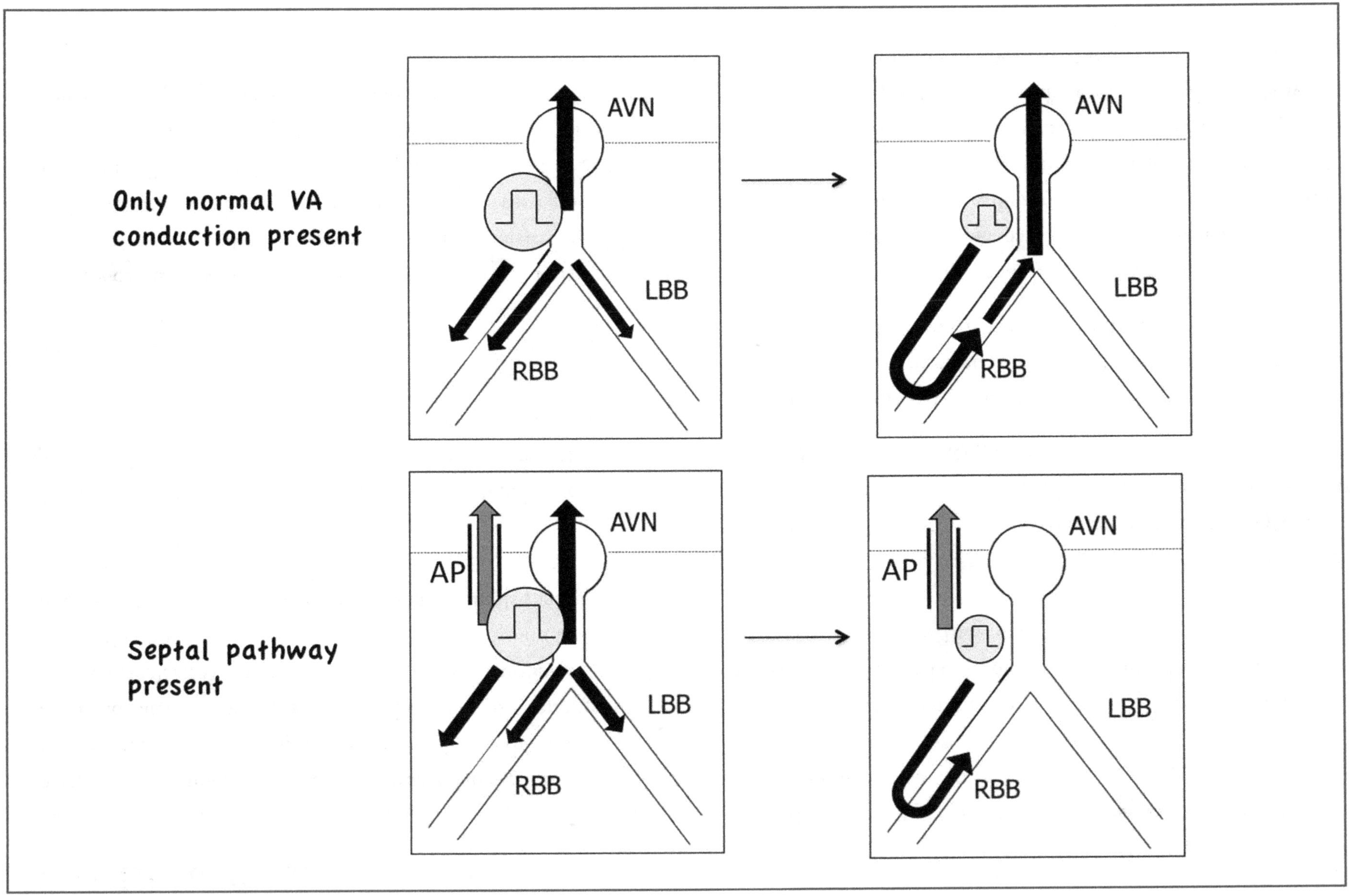
Only normal VA
conduction present
Septal pathway
present
AVN
LBB
RBB
AP

4. Loss of His bundle capture (smaller square wave symbols) in the *absence* of an AP (upper panel of Figure 2.2) will delay arrival of the impulse to the atrium (prolonging VA interval) since conduction needs to proceed over the ventricular myocardium from the para-Hisian region and return to the His bundle via the distal bundle branches. It will frequently also result in *delay* of arrival of the ventricular impulse at the apex, which is more closely related to the terminus of the RBB. The S-V interval at the His pacing catheter will not change and will remain short.

5. Loss of His bundle capture (smaller square wave symbols) in the *presence* of a septal AP close to the His bundle (lower panel of Figure 2.2) will *not* delay arrival of the impulse to the atrium, and hence no change in the VA interval. It will frequently *delay* arrival of the ventricular impulse at the apex, which is more closely related to the terminus of the RBB.

6. The further the AP is from the His bundle, the more the VA conduction time will prolong with loss of His capture. Thus, the maneuver is most useful for AP relatively closer to the His bundle.

A typical example of an expected response in the absence of an AP is shown in **Figure 2.3**. High output pacing results in a narrow QRS morphology suggesting capture of the bundle of His. The pacing output is gradually reduced and widening of the QRS morphology is noted (asterisk) signifying loss of His capture. The VA interval during His capture (70 ms) is much shorter than the VA interval during only para-Hisian ventricular capture (150 ms). This response is often referred to as a "nodal response" but is better thought of as indicating the absence of a septal AP. For example, one would expect prolongation of the VA time in the presence of an AP farther from the His bundle such as a lateral AP, the prolongation being in proportion to the distance of the AP from the His bundle.

Para-Hisian pacing in the presence of a septal AP is illustrated in **Figure 2.4**. The loss of His capture is indicated by sudden widening of the QRS complex. No substantive change occurs in the VA interval (170 ms as recorded at CS 1–2). Note also the change in morphology at the RVA EGM with a slight delay of the rapid component and prolongation of the S-V interval as would be expected with loss of His capture. This latter observation is not invariably present because of variability in the positioning of the RVA catheter but should be seen if the catheter is positioned near the terminus of the RBB.

**Figure 2.3**

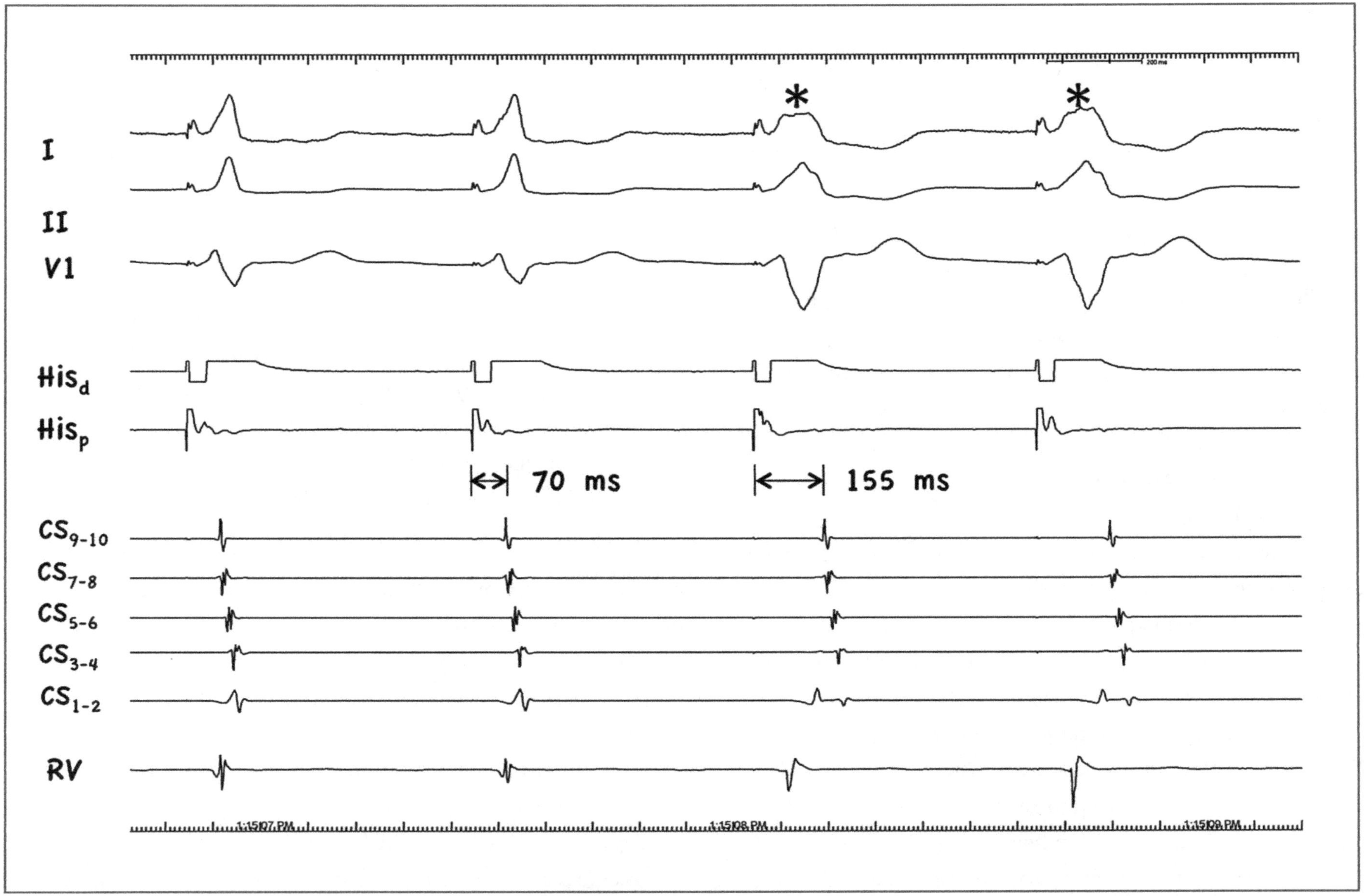

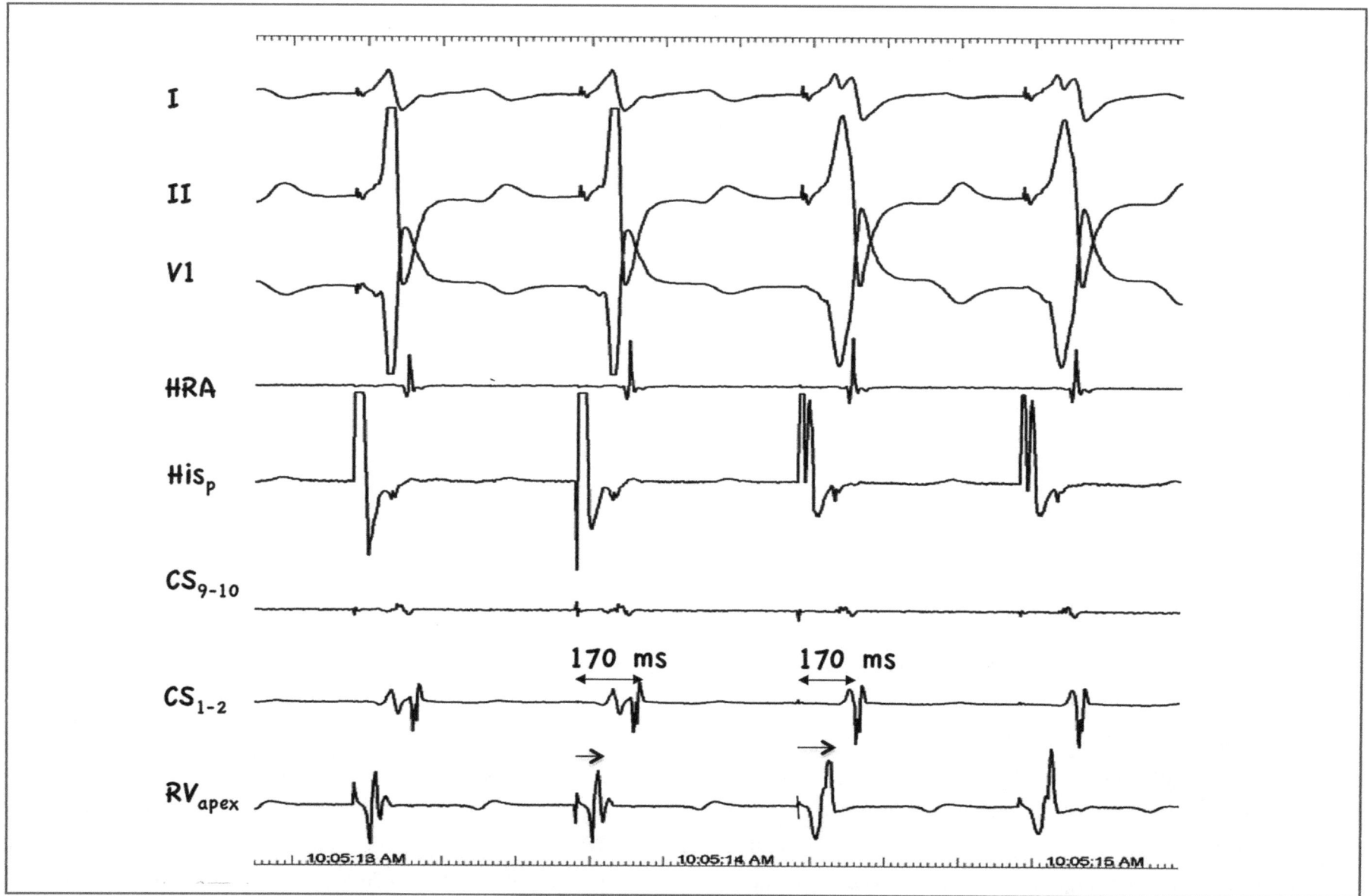

Figure 2.4

How would one interpret loss of His pacing in **Figure 2.5**? There is clearly loss of direct His capture in the 4th cycle. There is no change in the S-V at the His EGM. There is a slight change in the morphology of the V EGM at the RV$_{apex,}$ and the S-V concurrently prolongs as would be expected with loss of direct His pacing. There is no change in the VA interval, suggesting the presence of a septal AP. However, the VA interval of 55 ms at the high right atrium (HRA) is very short, and the same or slightly less at the His and CS 9–10. Short VA intervals in the range of 90 ms or less at the HRA and 60 ms or less at the CS orifice should raise the suspicion of inadvertent atrial capture, as was the case in this example. Slight displacement of the catheter distally, as shown in **Figure 2.6**, resulted in loss of atrial capture and demonstrated prolongation of the VA with loss of His capture supporting absence of a septal AP, as was the case. Inadvertent atrial pacing is a relatively common "pitfall" among those less experienced with this technique.

**Figure 2.5**

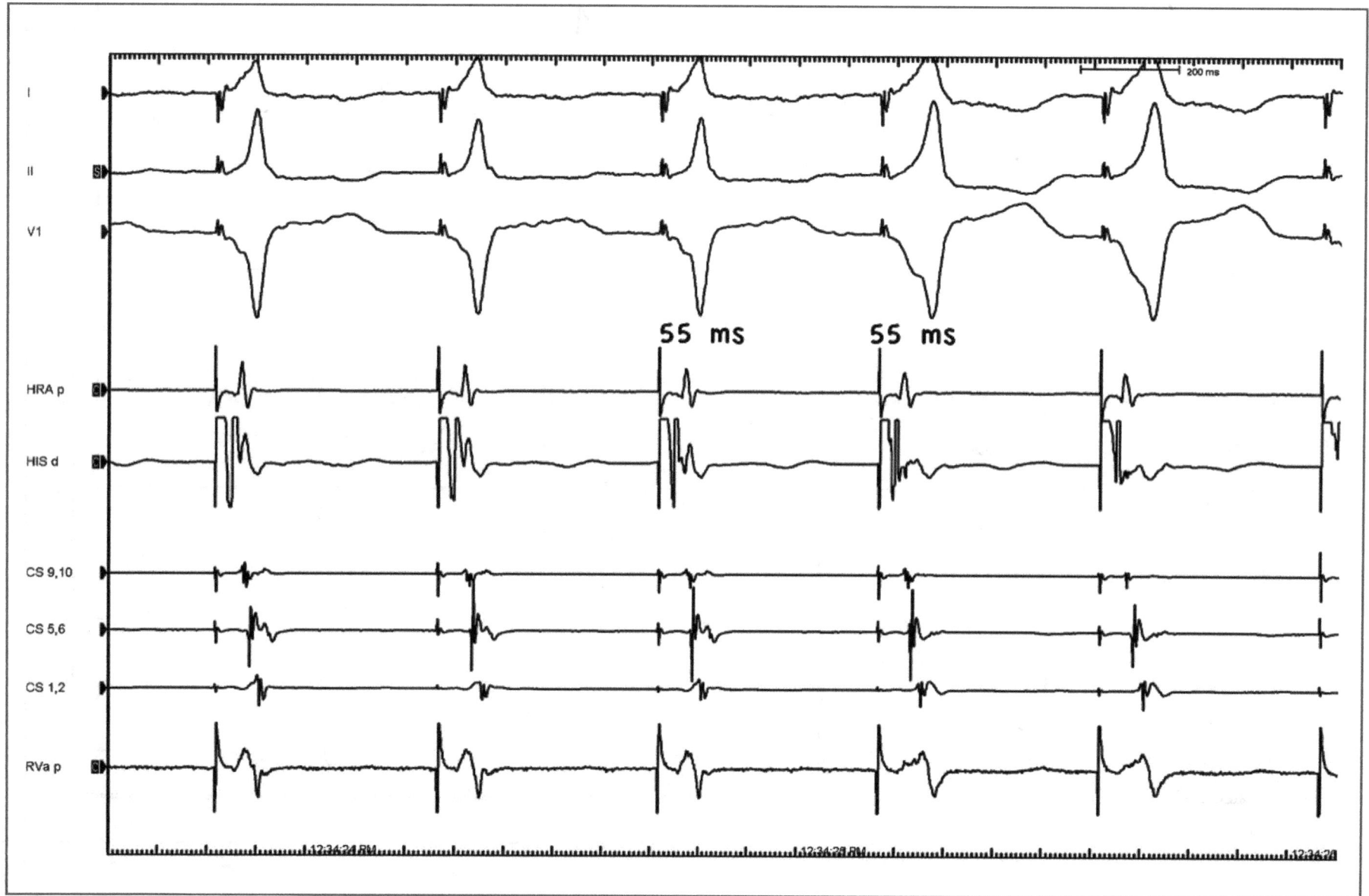

Figure 2.6

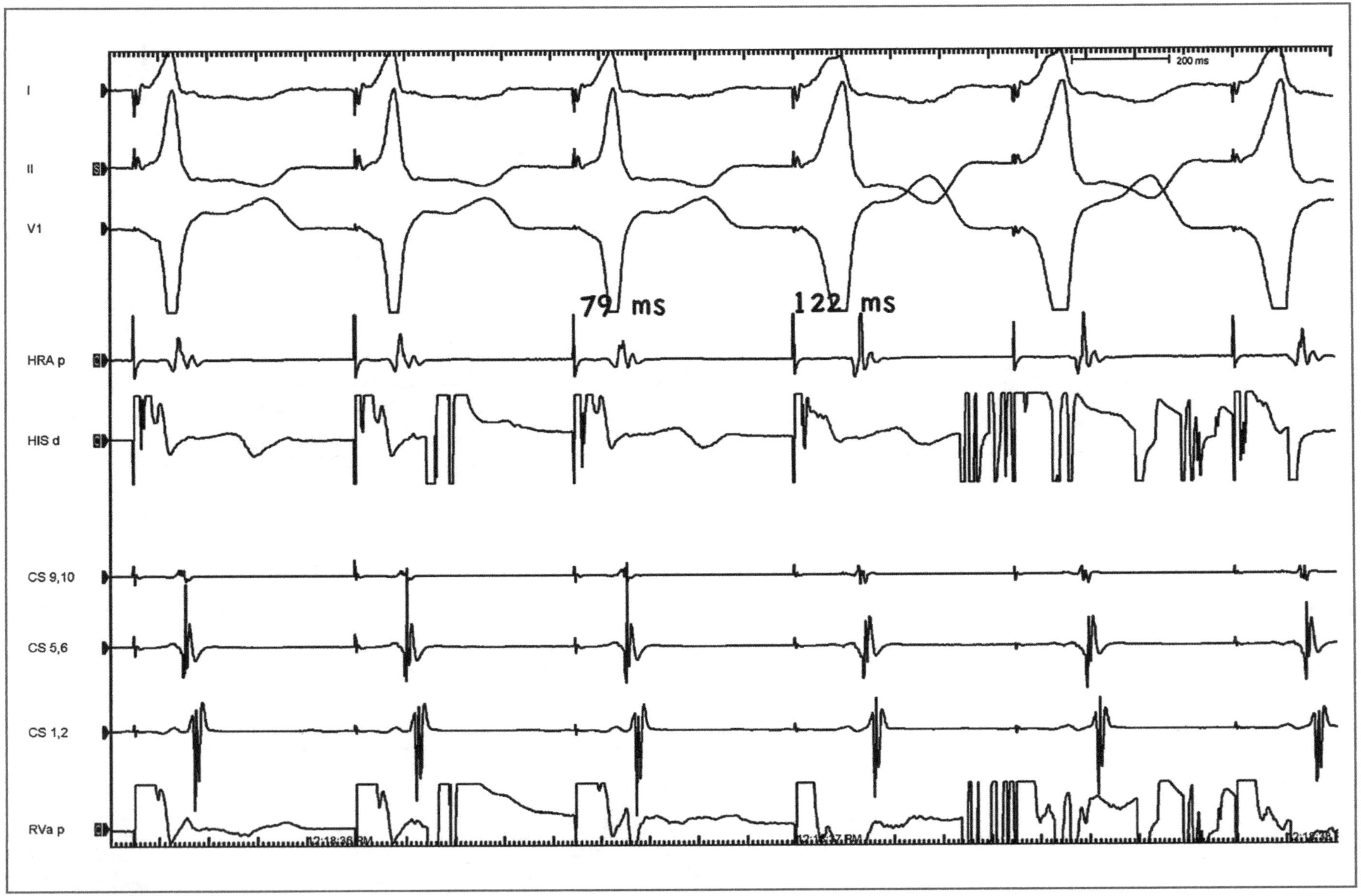

A limited lead array is shown in **Figure 2.7** to help make a point. One might consider that the CS could not be entered at EP study and one is still undecided about the mechanism of tachycardia. It is noted that the VA prolongs by 25 ms with loss of direct His pacing noted in the third cycle. Is this a "nodal" response? A better interpretation would be that one has disproved the presence of a septal AP (*with* the usual provisos that the conduction time over the AP is not longer than that over the normal AV conduction system and the AP is able to conduct at the pacing cycle length). The wisdom of this interpretation is apparent from **Figure 2.8**, where all the leads are shown and which clearly shows retrograde conduction over a left lateral AP. Loss of direct His pacing is verified by the emergence of the retrograde His from the ventricular EGM in the third cycle (small arrow).

**Figure 2.7**

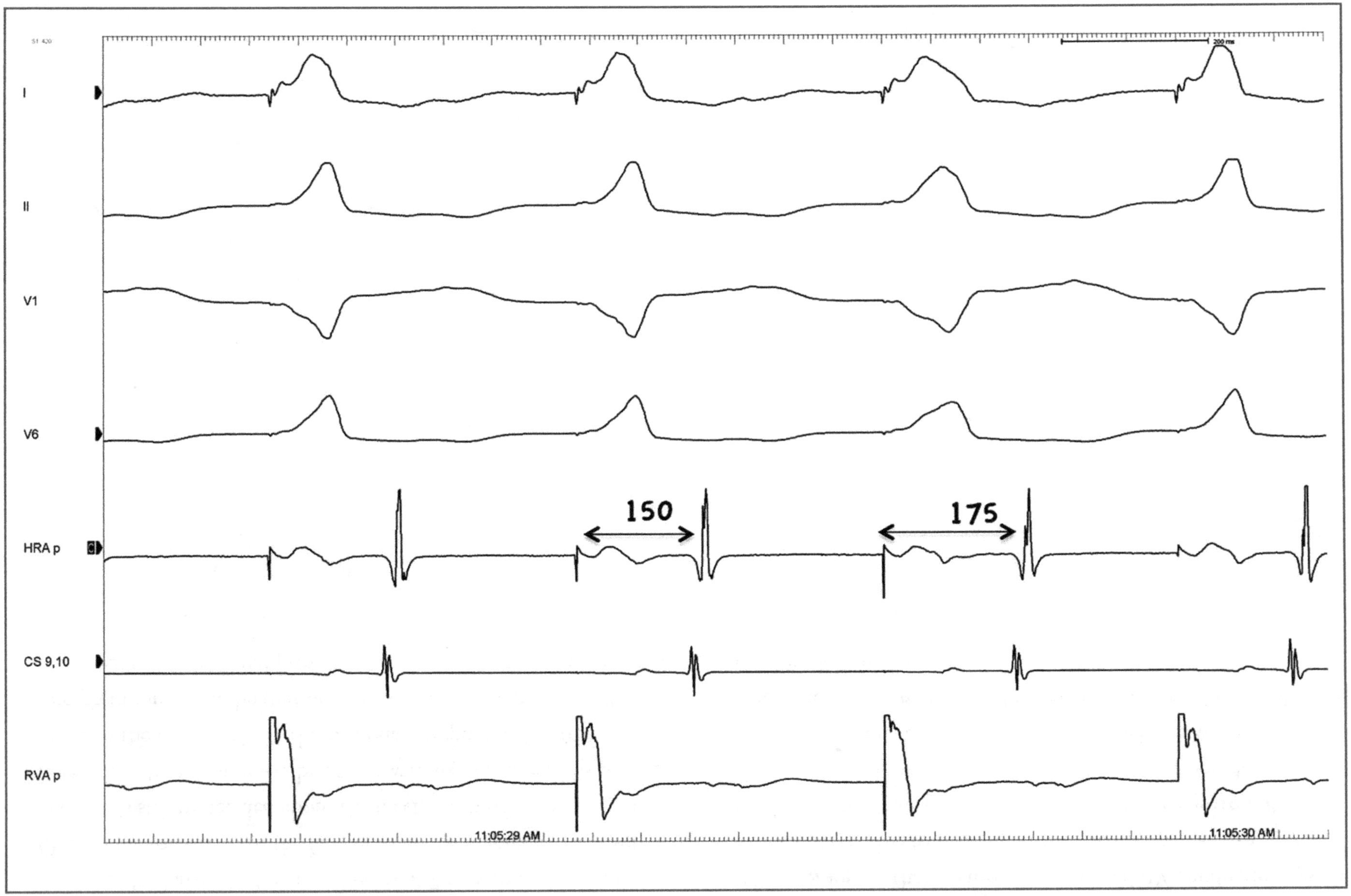

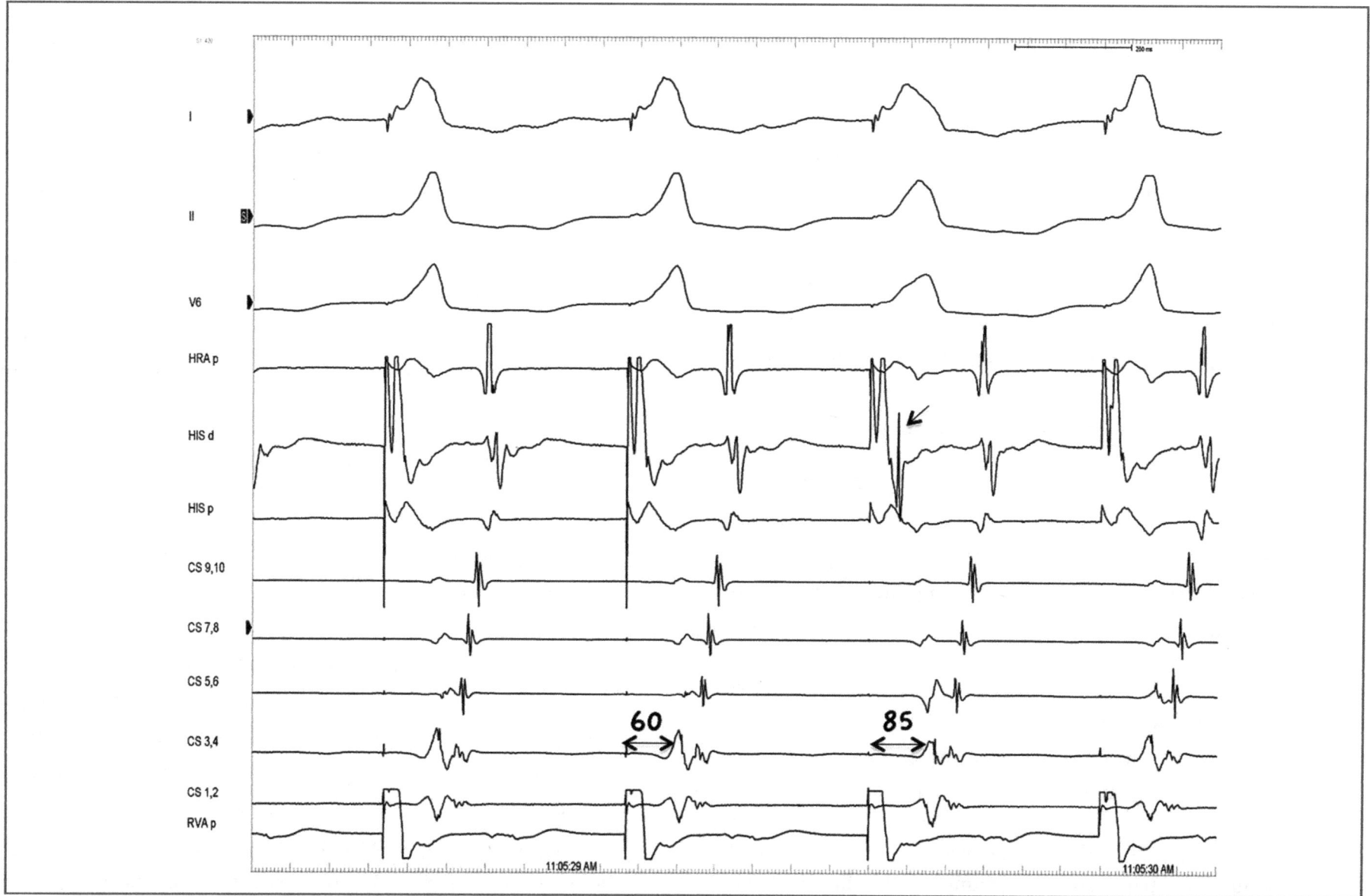

**Figure 2.8**

The patient represented in the tracing in **Figure 2.9** was referred for ablation of SVT with a central atrial activation. The third QRS shows loss of direct His pacing, verified by delay in activation of the His bundle (arrow). This is a very typical "AV nodal" response, which nonetheless should be interpreted with caution (see checklist below). **Figure 2.10** verifies that the clinical tachycardia was related to a septal AP with decremental conduction. The premature ventricular contraction (PVC) is "His refractory" and delays the subsequent A (reproducibly), proving the participation of the AP in the observed tachycardia. The "nodal" response observed during the para-Hisian maneuver was seen because conduction time over the accessory pathway was longer than that observed over the normal AVCS.

**Figure 2.11** is recorded from a patient with preexcitation proved to be related to a fasciculoventricular AP (inset). Para-Hisian pacing (first QRS) reproduces the preexcited QRS seen during sinus rhythm (last QRS), as one would expect with such an AP.

The following checklist may be useful when interpreting the response to the para-Hisian pacing maneuver as follows:

1. Verify that the atrium is not captured during pacing. A very short VA in the range of 60 ms or less at the CS orifice or 90 ms at the high right atrium should raise a red flag. A minor catheter repositioning generally corrects this.

2. The QRS during para-Hisian pacing should be relatively narrow and intermediate between the intrinsic normal QRS and the pure ventricular pacing QRS from the para-Hisian region.

3. A relatively early S-$V_{local}$ at the pacing catheter is expected with the capture of para-Hisian ventricular muscle. A relatively early S-$V_{local}$ at the RVA catheter is expected with capture of para-Hisian ventricular muscle (if properly positioned near the terminus of the RBB).

4. A QRS identical to the normal, nonpaced QRS should alert to the rare occurrence of direct His pacing without capture of the para-Hisian muscle. In this case, the S-$V_{local}$ at the His catheter should be relatively long and the S-$V_{local}$ at the apex should be relatively earlier.

5. Consider that a so-called "AV nodal" response may mean that a lateral AP is present, an AP is present but has a long conduction time, or AP is present but is refractory at the cycle length used for pacing. A septal AP response suggests but does not prove that the AP is involved in the clinical tachycardia.

Figure 2.9

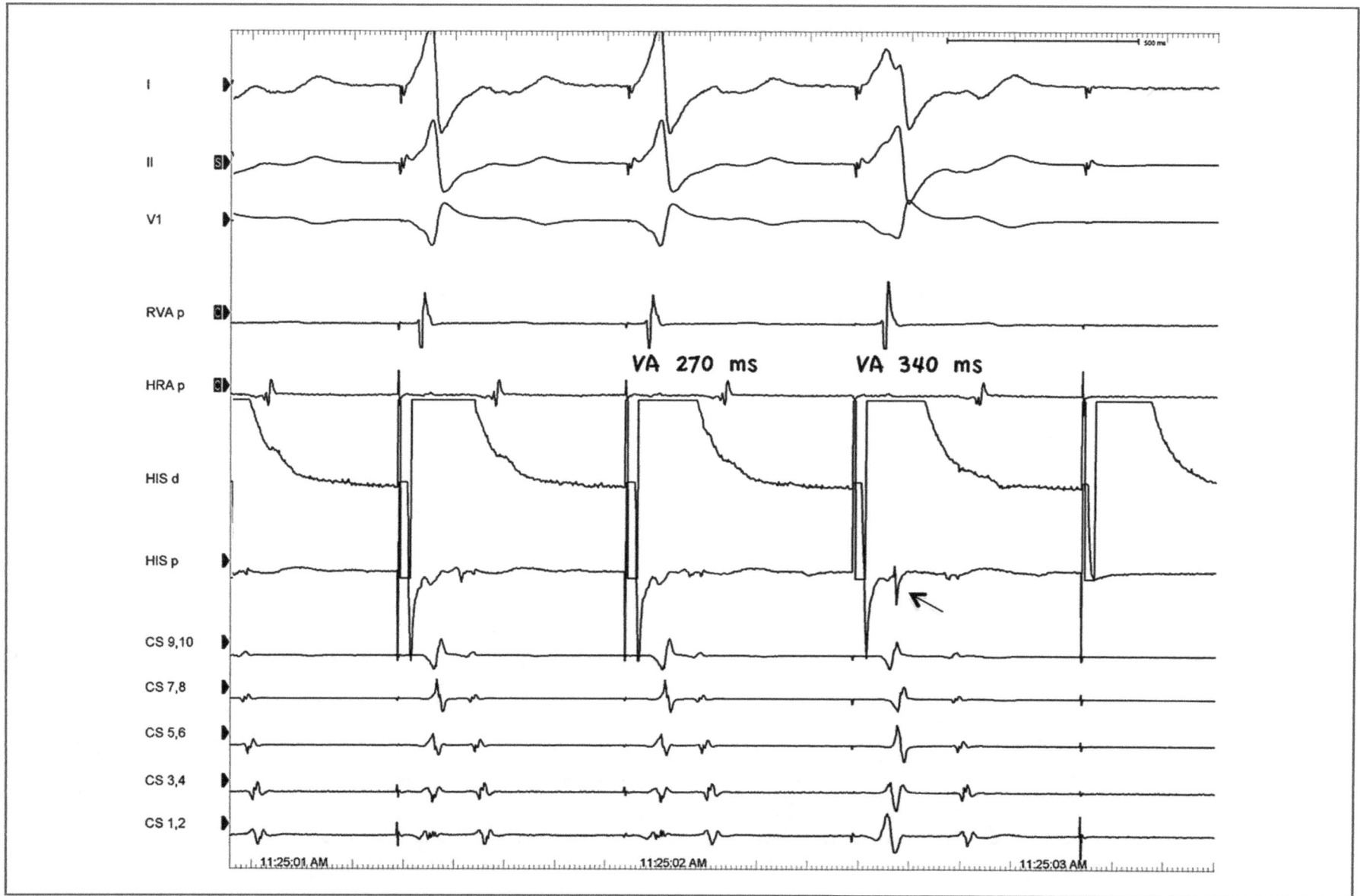

Figure 2.10

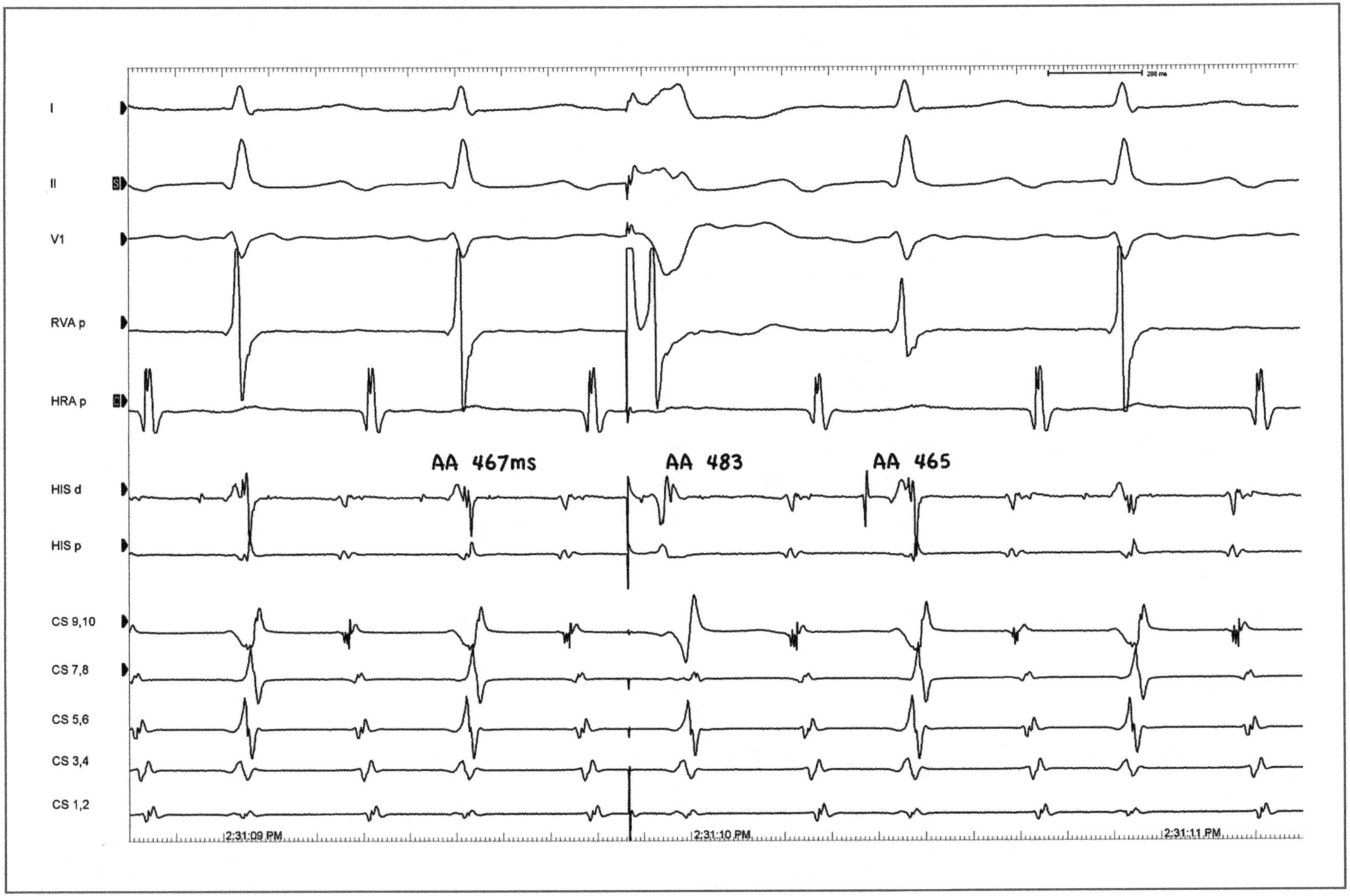
I
II
V1
RVA p
HRA p
AA 467ms
AA 483
AA 465
HIS d
HIS p
CS 9,10
CS 7,8
CS 5,6
CS 3,4
CS 1,2
200 ms
2:31:09 PM
2:31:10 PM
2:31:11 PM

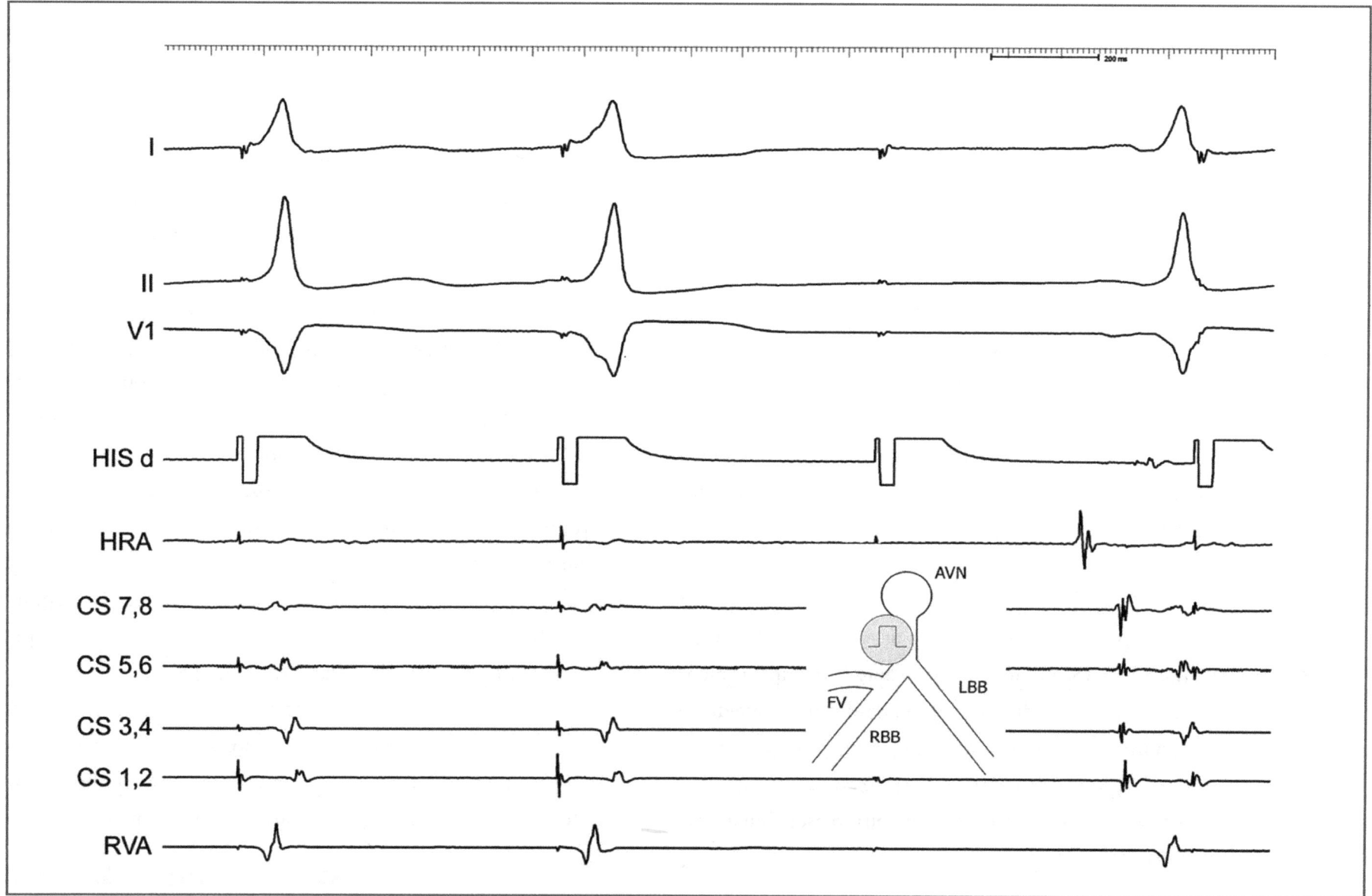

Figure 2.11

# Assessing a Line of Block

Large reentrant circuits frequently utilize "channels" bordered by anatomic boundaries, and creation of block in these channels by ablation is done in various contexts in clinical EP. Some method to assess completeness of the line of block is necessary after ablation. There are many reported methods, including looking for widely split potentials along the ablation line and assessing activation sequence with multipolar catheters during pacing at one side of the line. A simple pacing maneuver involving pacing at variable distances on one side of the line and measuring the arrival of the impulse on the other (called "differential pacing" by some) is widely used for this, and the principle remains the same whether one is assessing a tricuspid-caval line, a left atrial (LA) roof line, or a mitral isthmus line. Of course, no single method is infallible, as ultimately conduction may be slowed to the point where it is indistinguishable from irreversible block only to recover with time.

This principle is illustrated in **Figure 2.12** for a "generic" line of block hypothetically just created by catheter ablation (green line) between 2 anatomic barriers (black rectangles). Consider that you are pacing close to the line at $P_1$ and $P_2$, respectively, and recording from the other side of the line (R). It is then obvious that the time to get from $P_1$ to R (i.e., $P_1$–R) is greater than the time $P_2$–R when complete block is present and the impulse needs to take the "long route" to the other side of the line of block (left panel). Conversely, when a gap is present in the line (right panel), $P_1$–R will be shorter than $P_2$–R.

Since block may be unidirectional, the exercise is then repeated reversing the pacing and recording sites to verify that block is demonstrable by pacing in either direction.

It is also important to pace and record as close as possible to the line when doing this maneuver to minimize the possibility of myocardial conduction shunting to the other side by an alternate route.

The following will include applications of differential pacing at specific sites.

**Figure 2.12**

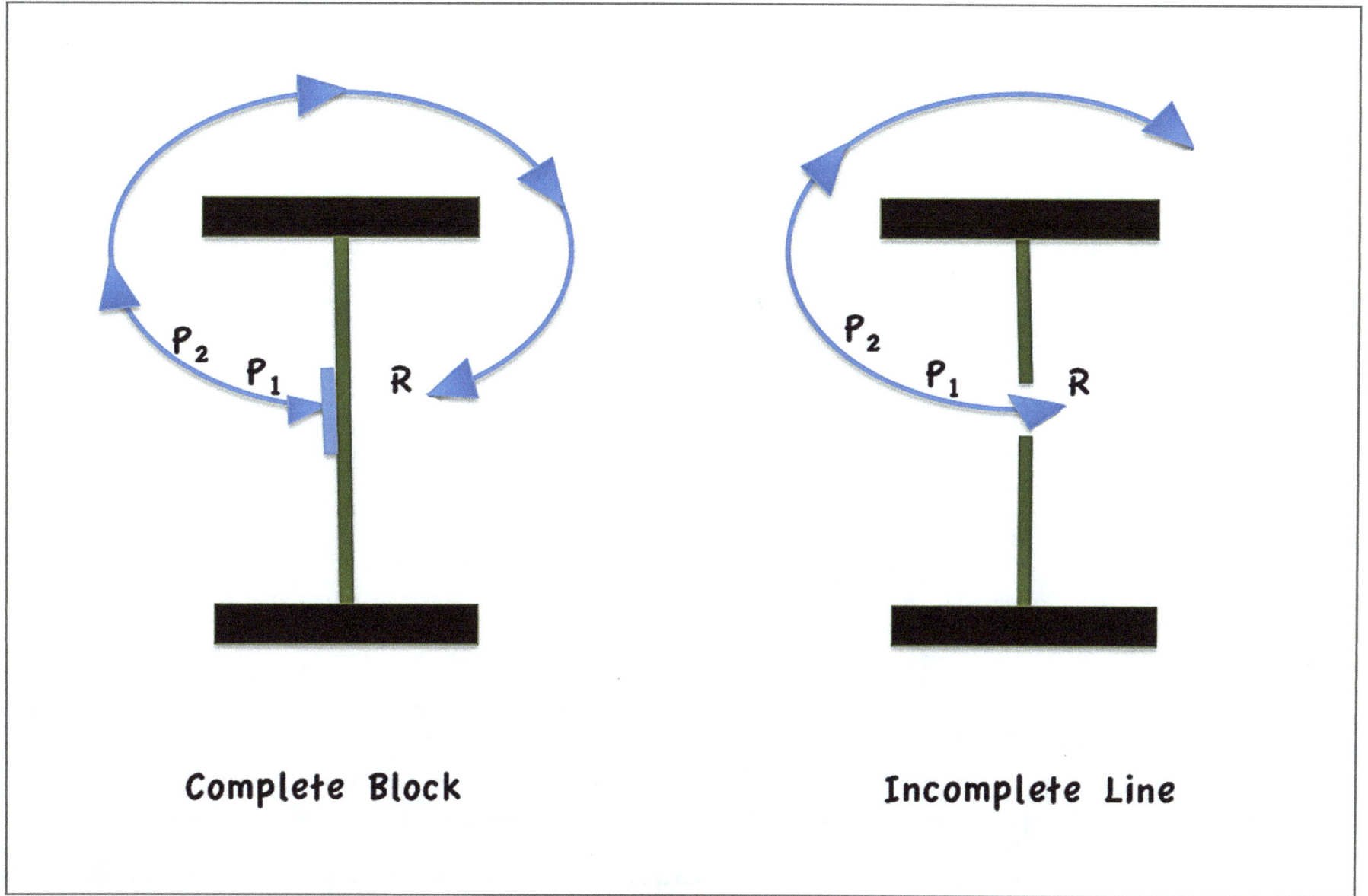

Complete Block                    Incomplete Line

## Assessing a Line of Block: Tricuspid-Caval Isthmus

One of the most common clinical procedures is ablation of the tricuspid-caval isthmus for "isthmus-dependent" atrial flutter. After ablation along the line, one might observe split potentials along the line and perhaps change of activation sequence with a multipolar circular catheter if such is used.

The differential pacing maneuver can then be performed as illustrated. The larger circle represents the tricuspid valve annulus (TVA) and the double line represents the line of block. In **Figure 2.13**, the proximal CS is paced initially while recording close

to the line at site (A). The time CS-A (second circle) is measured (182 ms). Moving the recording site slightly more lateral of the line to site (B) results in shortening of conduction, with CS-B (third circle) = 162 ms.

The maneuver is performed in reverse in **Figure 2.14**, where pacing the other side closer to the line (A-CS = 182 ms) results in a longer interval than pacing farther (more lateral) from the line (B-CS = 162 ms).

Figure 2.13

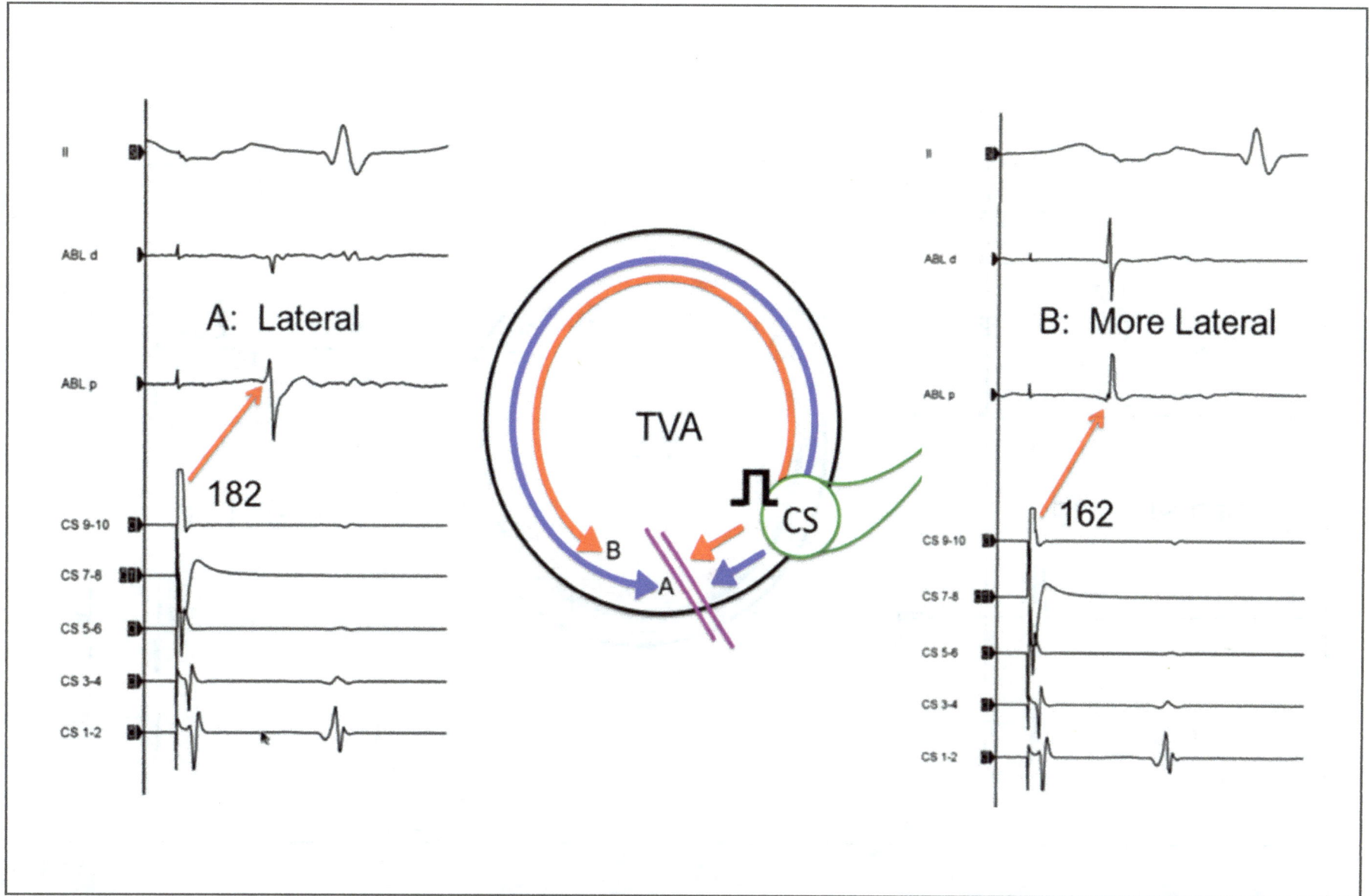

Figure 2.14

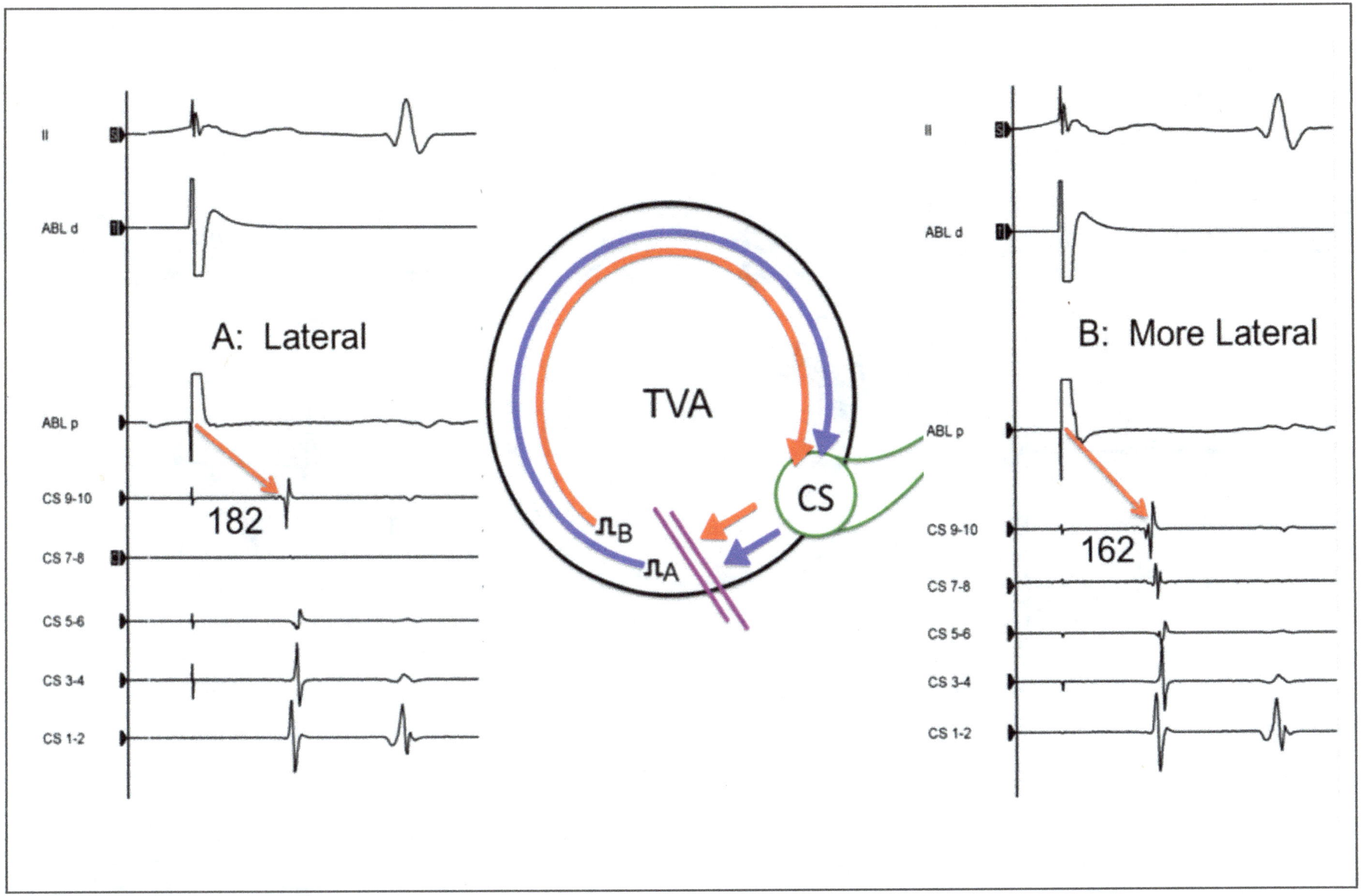

II
ABL d
A: Lateral
ABL p
CS 9-10
182
CS 7-8
CS 5-6
CS 3-4
CS 1-2
TVA
CS
B
A
II
ABL d
B: More Lateral
ABL p
CS 9-10
162
CS 7-8
CS 5-6
CS 3-4
CS 1-2

# Assessing a Line of Block: Left Atrial Roof Line

LA roof-dependent flutter can occur following pulmonary vein (PV) isolation and is managed by creating a line of block in the superior LA joining the circumferential lines of block around the PVs as depicted in **Figure 2.15**. This and subsequent similar figures depict the left atrium as viewed from the front with anterior wall removed.

The LA appendage is anterior to the roof line and paced ("LAA"). EGM measurements can be made at points 1 and 2 on the posterior wall. If the roof line is complete, time LAA to recording point 1 at the back of the LA should be longer than point 2, i.e., Stim LAA-1 > Stim LAA-2.

To test anterior conduction block across the roof line, point 1 can be paced and compared to pacing at point 2. If there is anterior conduction block, the Stim-EGM interval should be longer pacing from point 1 compared to point 2.

From the schematic in the figure, *a potential problem of pacing relatively far from the roof line is evident*. For example, if the pacing catheter is situated in a low LA appendage site (B), then the Stim-EGM may be longer at point 1 compared to point 2 *even without posterior conduction block at the roof*. One might also consider pacing closer to the line (site C). If the block is present, Stim LAA-1 < Stim C-1.

Mapping along the line of ablation in the example (**Figure 2.16**) was done during LAA pacing and demonstrates a complex EGM in the distal ablation channel (inset shows the magnified EGM).

The solid arrows show relatively low amplitude, probably far-field signals. The first far-field EGM arises from the LA roof anterior to the line (solid arrow). The second far-field EGM (solid arrow) arises from the LA posterior wall. Note a near-field signal of similar timing oriented in the posterior LA near the line.

The widely split double potential suggests block along the line but a high-frequency fractionated EGM (dashed arrow) suggests a gap in the partially ablated roof line. Ablation at this site delayed the second far-field EGM further and roof line block was verified as follows.

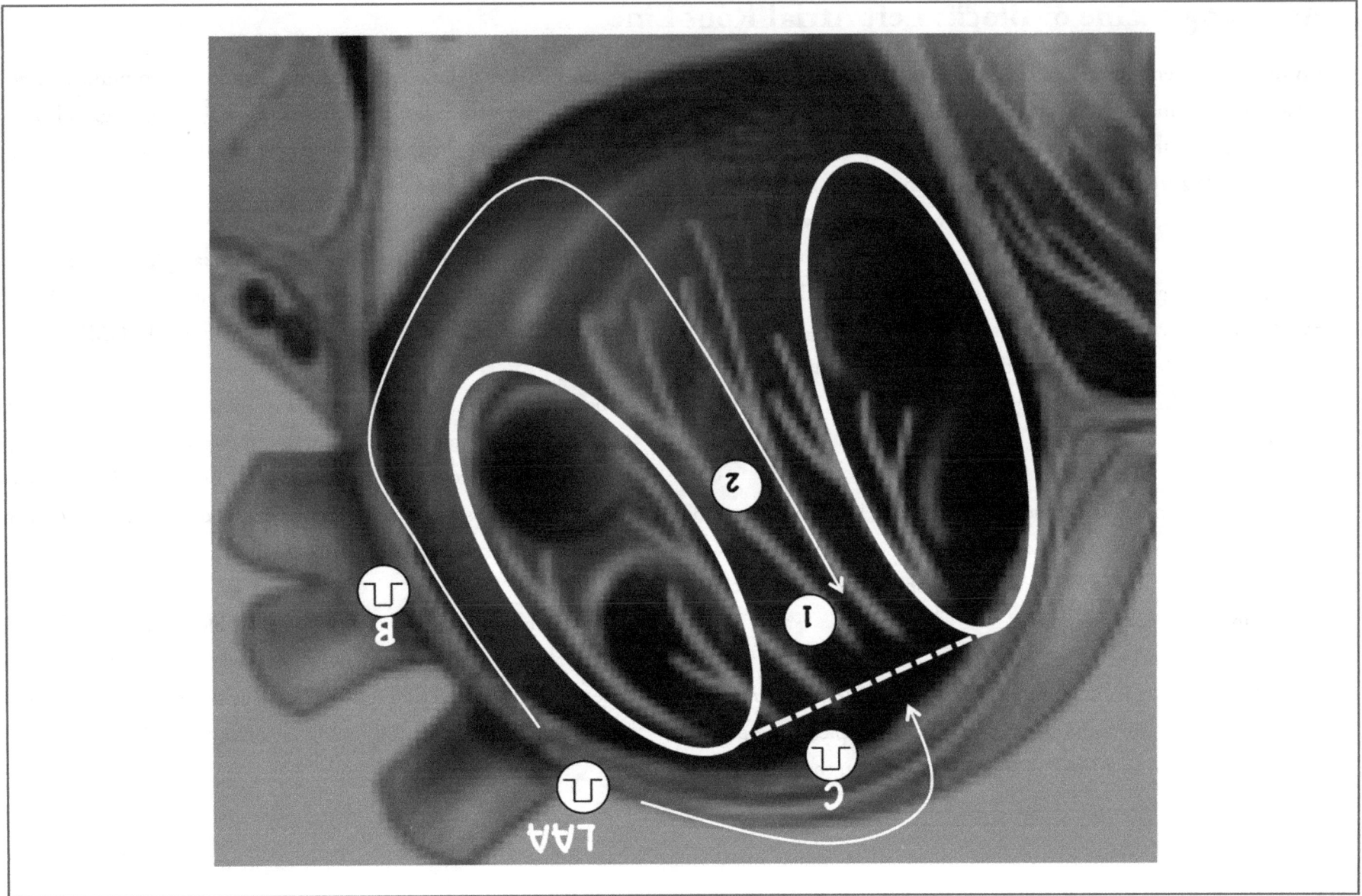

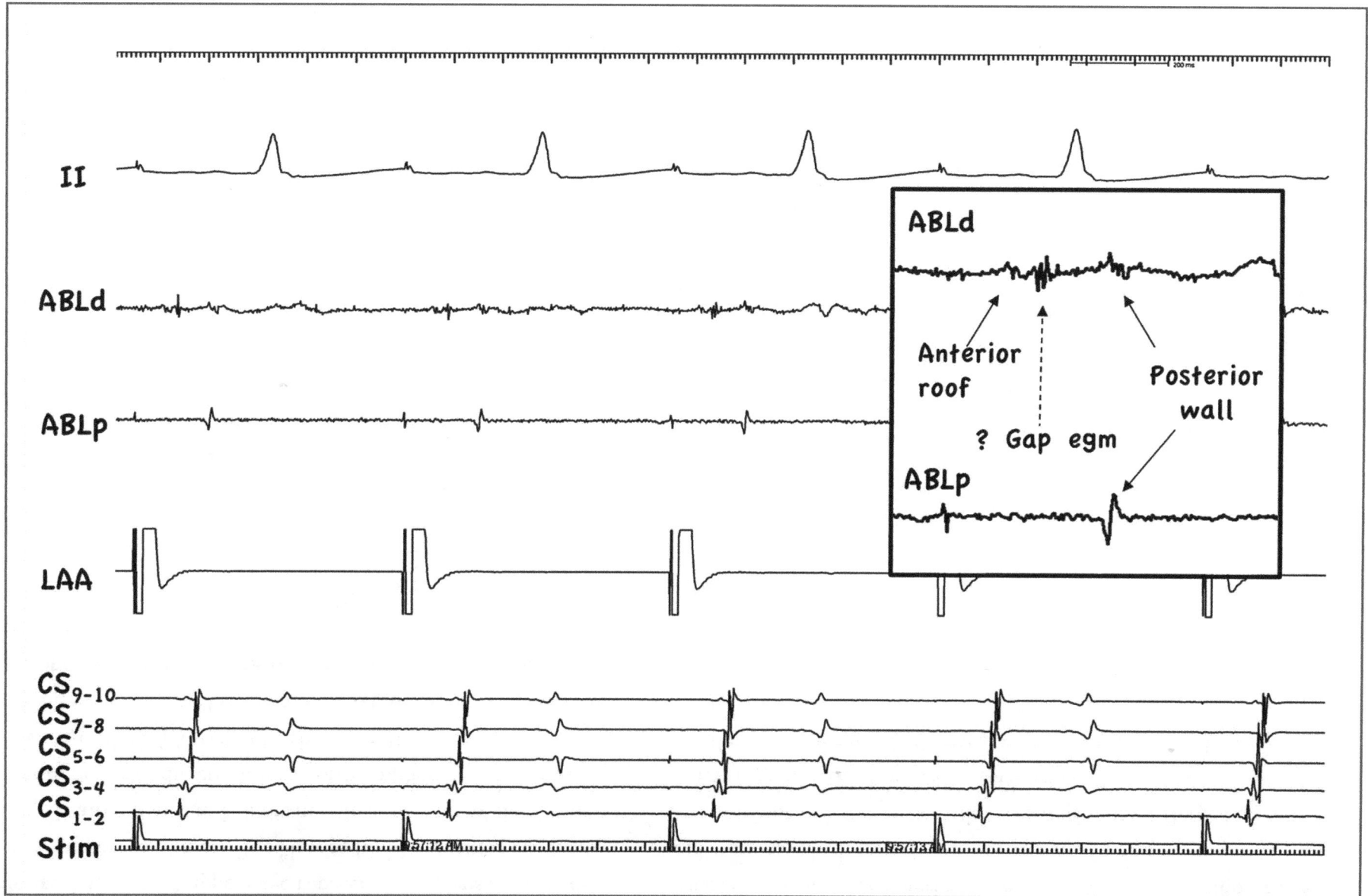

Figure 2.16

In **Figure 2.17**, the LA appendage is paced and signals are recorded on the posterior wall (inset). The "low to high" activation sequence (i.e., site 2 earlier than site 1) indicates block of the roof line in the anterior to posterior direction. It is worth emphasizing that too low positioning of the LAA pacing further from the ablation line makes the pacing site relatively closer to site 2 than site 1 and increases the chance of observing a false positive, i.e., apparent block when in fact not blocked.

Once the block in the anterior to posterior direction is demonstrated, bidirectional block is likely but not proven. To do so, one would pace the posterior wall at 2 sites while recording anterior to the line at the LAA (**Figure 2.18**). It is evident that Stim-LAA is longer at site 1 (140 ms) than at site 2 (120 ms) as one would expect with posterior to anterior block of the roof line.

Figure 2.17

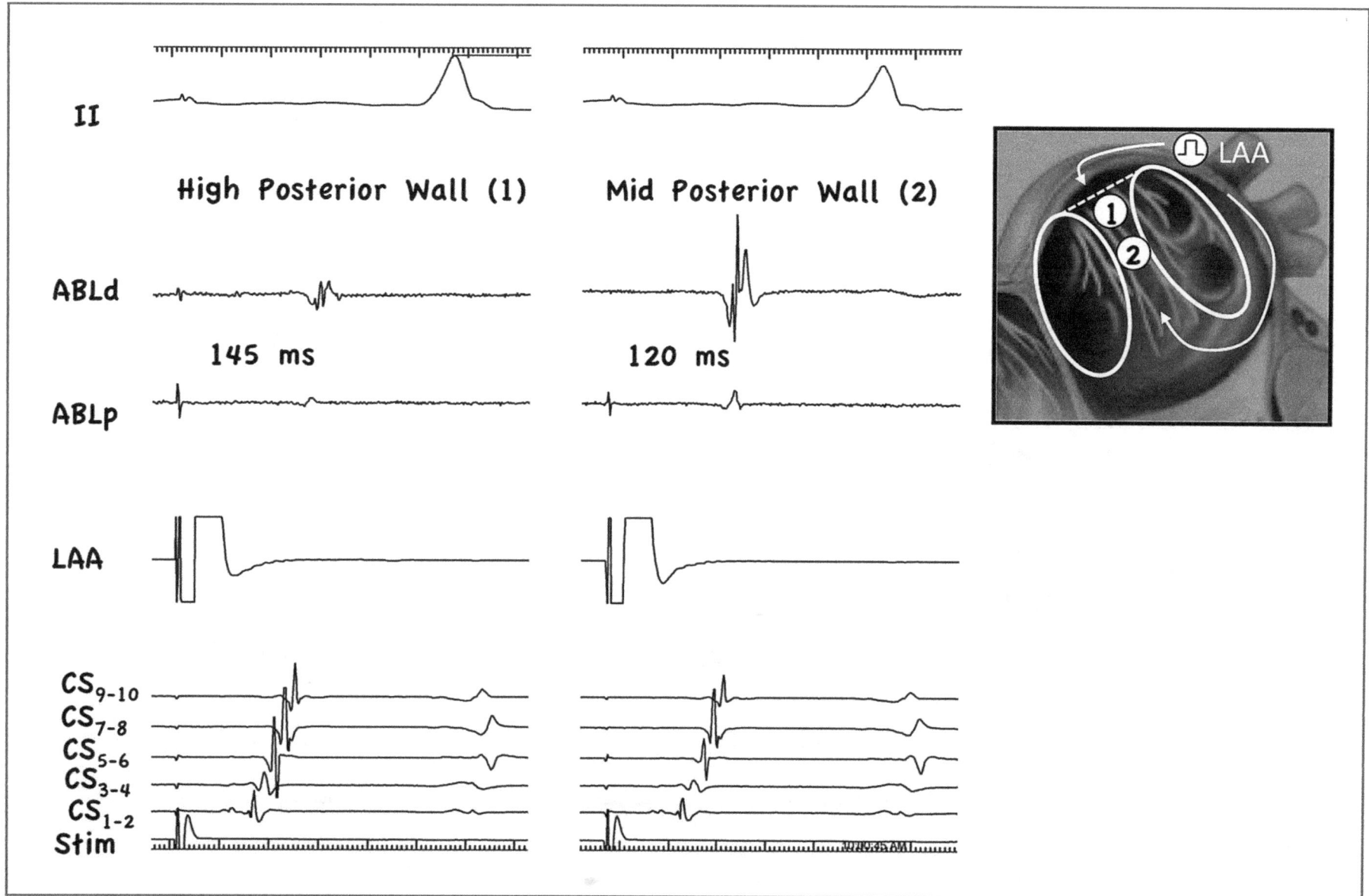

II

**High Posterior Wall (1)**    **Mid Posterior Wall (2)**

ABLd

145 ms    120 ms

ABLp

LAA

CS$_{9-10}$
CS$_{7-8}$
CS$_{5-6}$
CS$_{3-4}$
CS$_{1-2}$
Stim

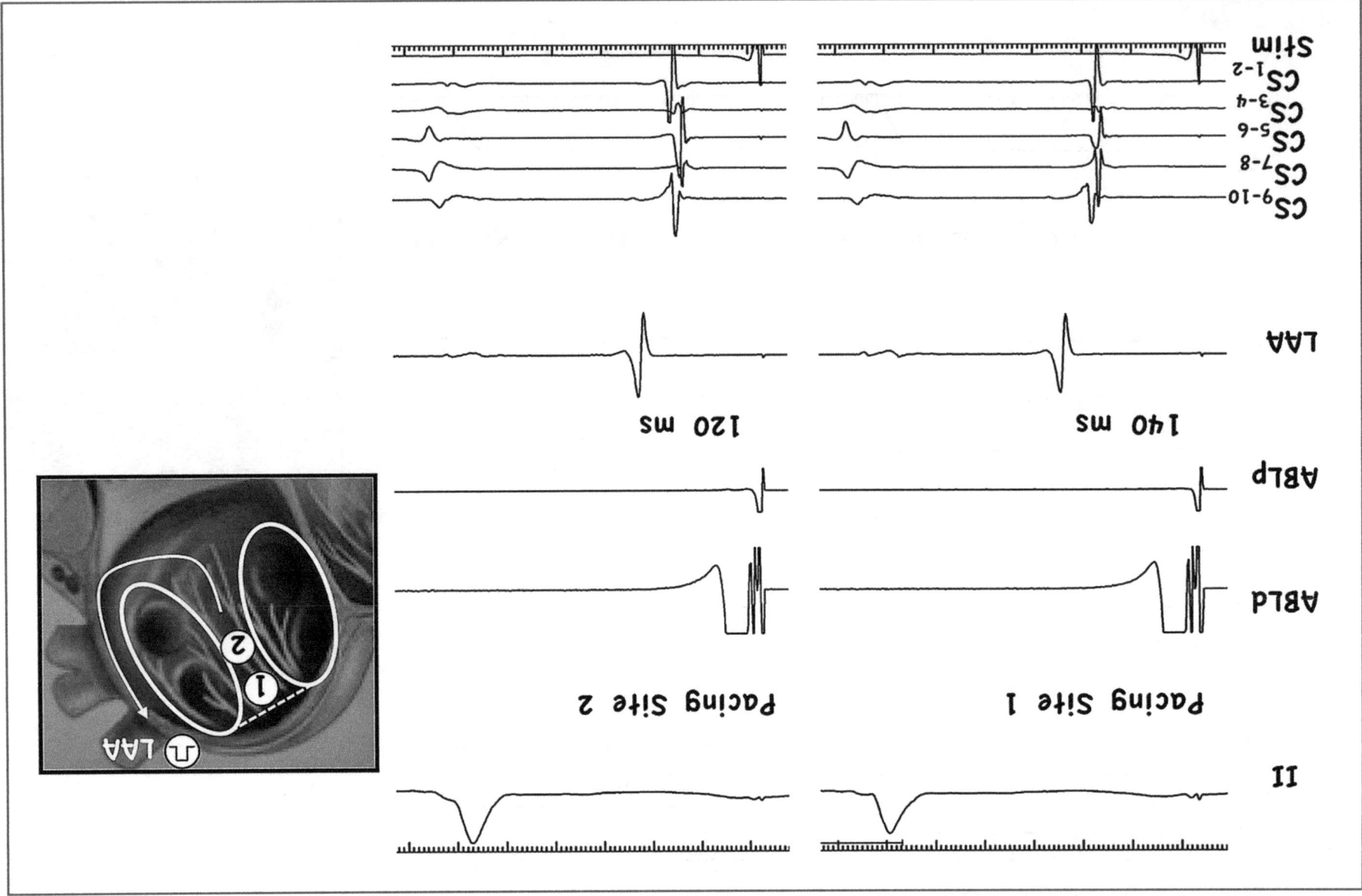

**Figure 2.18**

## Assessing a Line of Block: Mitral Isthmus Line

Mitral annular flutter generally occurs after pulmonary vein (PV) isolation or other cardiac surgery. To interrupt this circuit, the "isthmus" between the mitral valve and the inferior border of the antral circle around the PVs is generally targeted as most convenient with the usual site from the left lower PV to 4 o'clock on the mitral isthmus as illustrated in **Figure 2.19** (dashed line). The LA appendage (A) paces from a superior (or "lateral") direction to the mitral isthmus line. EGM measurements can be made at points 1 and 2 on the mitral isthmus, usually recorded from the CS catheter. If there is clockwise conduction block across the mitral isthmus line, the Stim-EGM interval should be longer at point 1 than at point 2.

To test counterclockwise conduction block, point 1 can be paced and compared to pacing at point 2. If there is counterclockwise conduction block, the Stim-EGM interval should be longer pacing from point 1 compared to point 2.

Emphasizing this important issue again, pacing too far from the line, such as from site C, may result in erroneous diagnosis of clockwise block if conduction to point 2 precedes conduction to point 1. Although the LA appendage is usually satisfactory and convenient, a site closer to the line, such as site B, would theoretically be even more rigorous.

In **Figure 2.20**, the LA appendage is paced during ablation at the mitral isthmus. The baseline CS activation during LA appendage pacing demonstrated a fused wave front, both proximal to distal and distal to proximal (solid arrows). During ablation, there is a sudden change in the Stim-EGM interval (asterisk) in the distal CS, whereas the proximal CS activation is unchanged. The activation is now completely proximal to distal (dashed arrow). This sudden change suggests mitral isthmus block but a differential pacing maneuver is required to prove it.

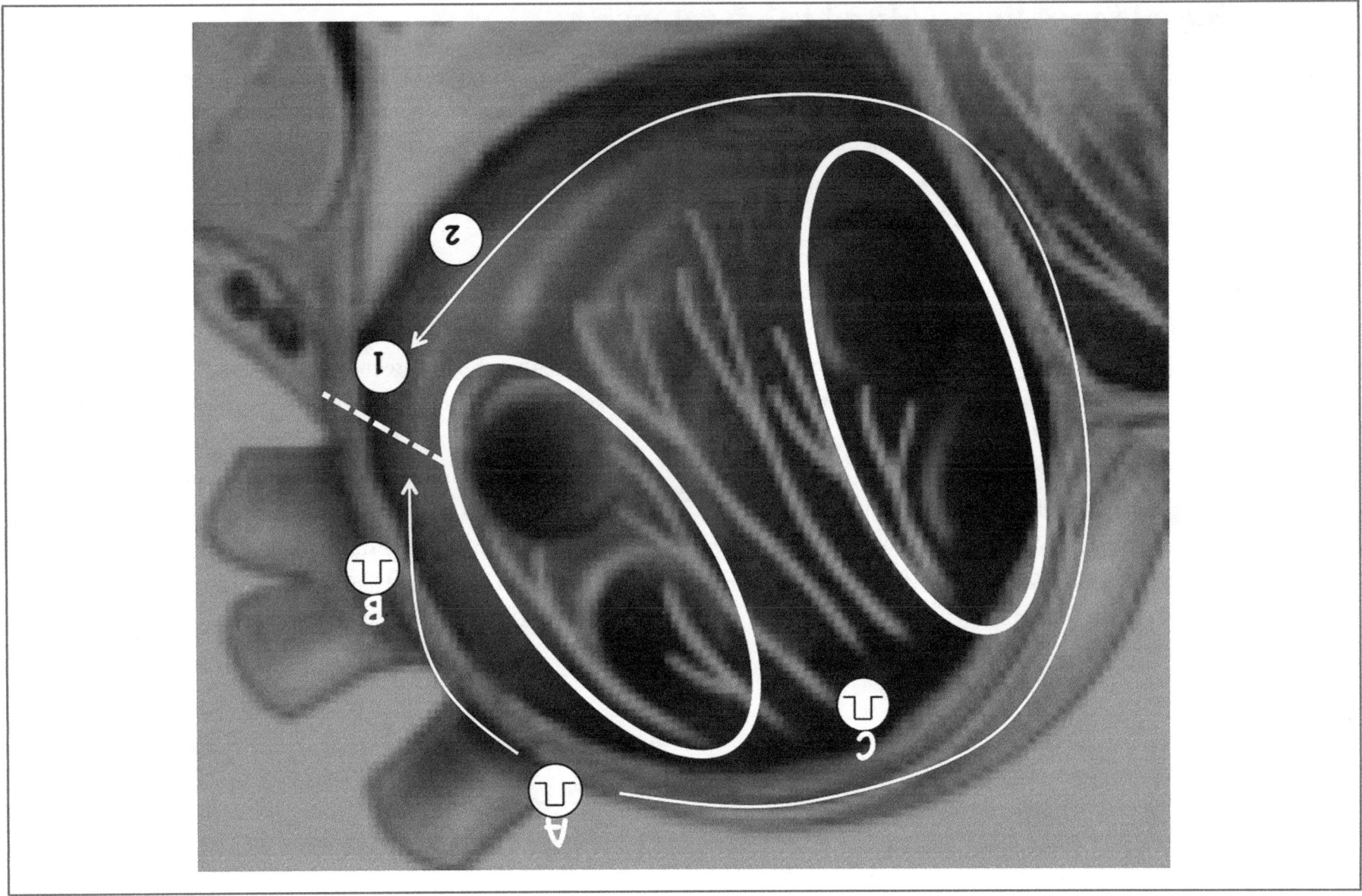

Figure 2.19

**Figure 2.20**

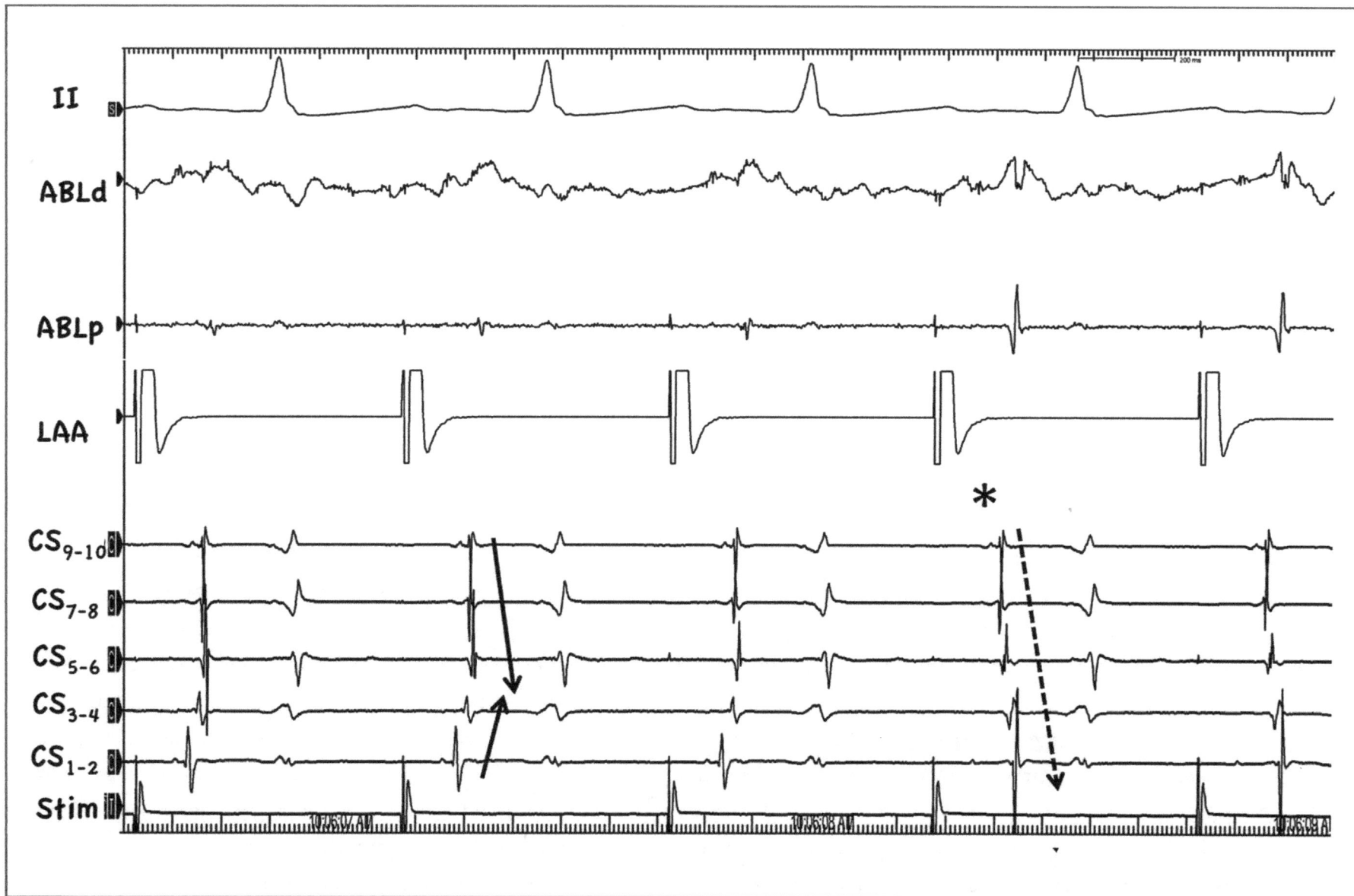

In **Figure 2.21**, clockwise mitral isthmus block is being evaluated. The LA appendage is paced and the first indication of clockwise block is the CS activation pattern, which is proximal to distal. This may be misleading depending on the position of the CS catheter relative to the ablation line. Therefore, the ablation catheter is initially placed just posterior to the mitral isthmus line (1). Placing the catheter more septally (2) demonstrates a shorter Stim-EGM interval (corresponding to the activation sequence seen on the CS catheter) and supporting clockwise mitral isthmus block.

To verify bidirectional block, counterclockwise conduction is evaluated as follows. The pacing catheter can be left in the LA appendage and used for recording. Pacing a site just posterior to the mitral isthmus line (1) should result in a longer Stim-EGM than a more septal pacing site (2). (**Figure 2.22**)

The CS catheter may be used for pacing, pacing a distal bipole, and comparing it to pacing at an adjacent, more proximal bipole. One needs to ensure that the distal bipole chosen is posterior to the ablated line and not superior to it.

**Figure 2.21**

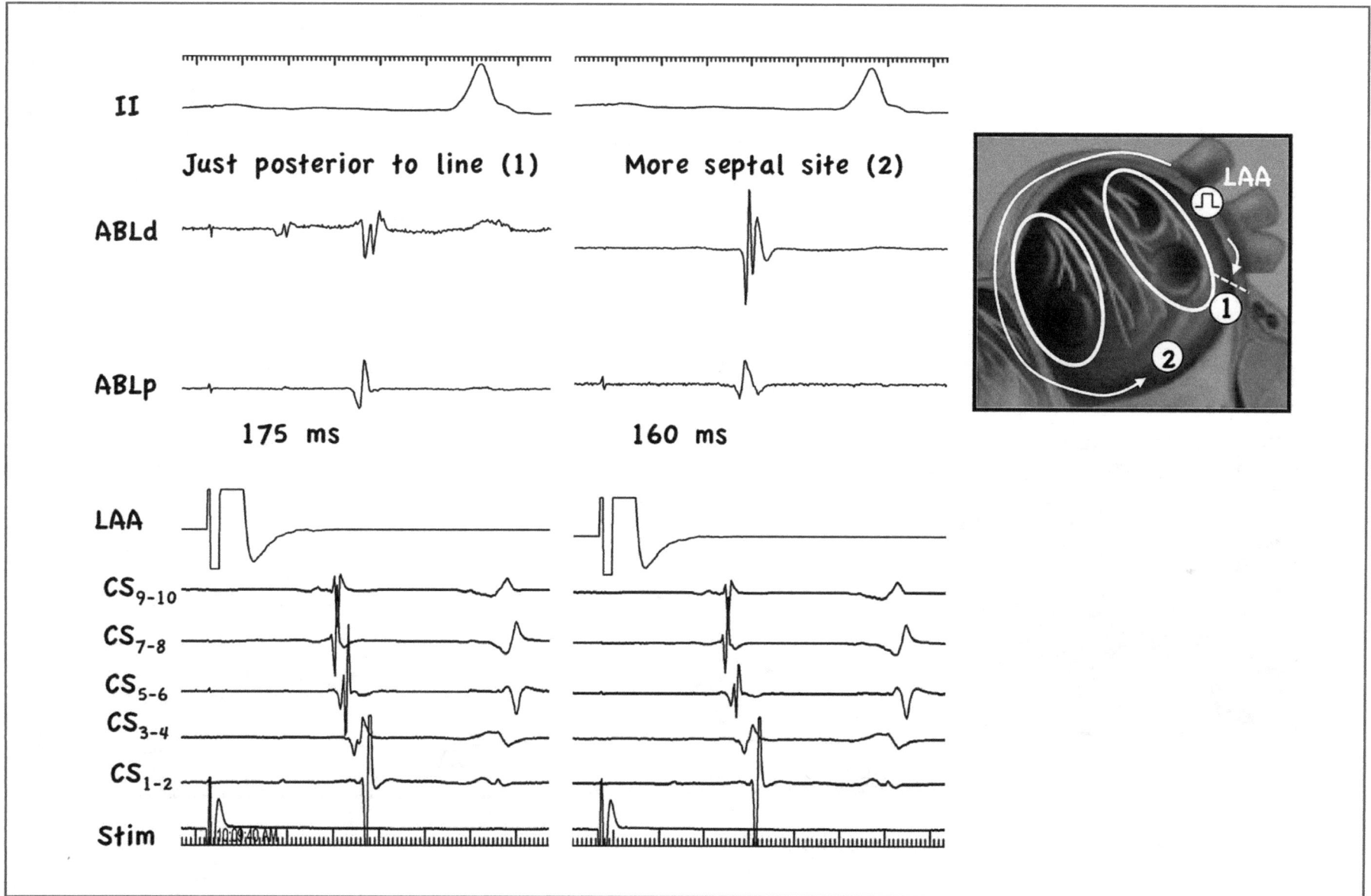

II

Just posterior to line (1)          More septal site (2)

ABLd

ABLp

175 ms          160 ms

LAA

CS₉₋₁₀

CS₇₋₈

CS₅₋₆

CS₃₋₄

CS₁₋₂

Stim

Figure 2.22

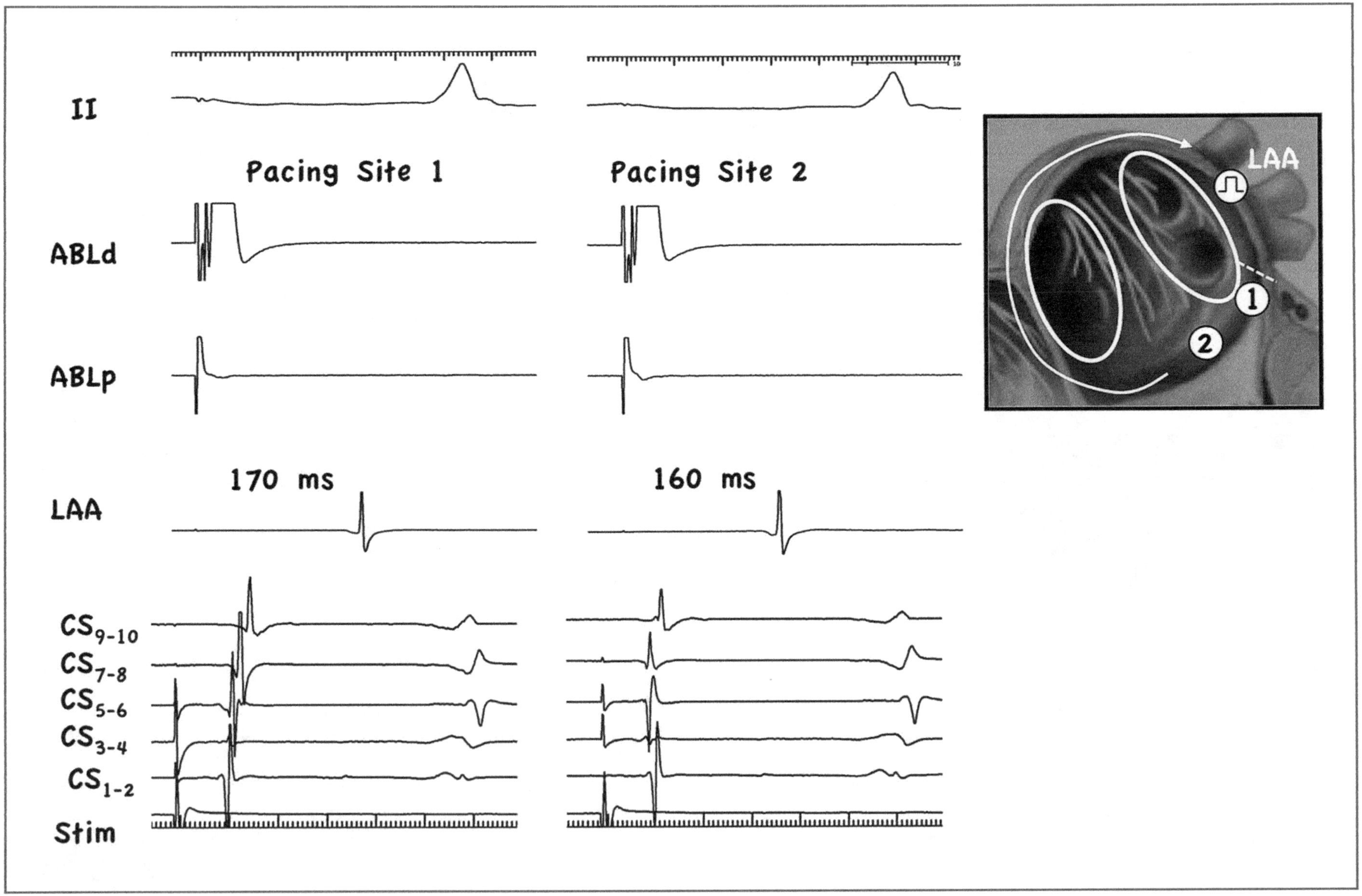

II
Pacing Site 1
Pacing Site 2
ABLd
ABLp
LAA
170 ms
160 ms
$CS_{9-10}$
$CS_{7-8}$
$CS_{5-6}$
$CS_{3-4}$
$CS_{1-2}$
Stim
LAA
1
2

# Verifying PV Isolation

Isolation of the PVs remains an important element of atrial fibrillation (AF) ablation. Entrance block to the PV is signaled (**Figure 2.23**) when PV potentials inside the PV (RSPV, arrows) disappear during ablation (asterisk). This usually but not invariably means bidirectional block. The latter can be confirmed by pacing inside the PV (**Figure 2.24**). Capture of the right superior PV is demonstrated in both the proximal ablation bipole (dashed arrow) and the multipolar catheter situated in the right superior PVs (solid arrow). Concurrently, the remainder of the atria are in sinus rhythm with the LA EGMs in the CS catheter dissociated from the pacing in the PV. Thus, exit block from the right superior PV is also demonstrated.

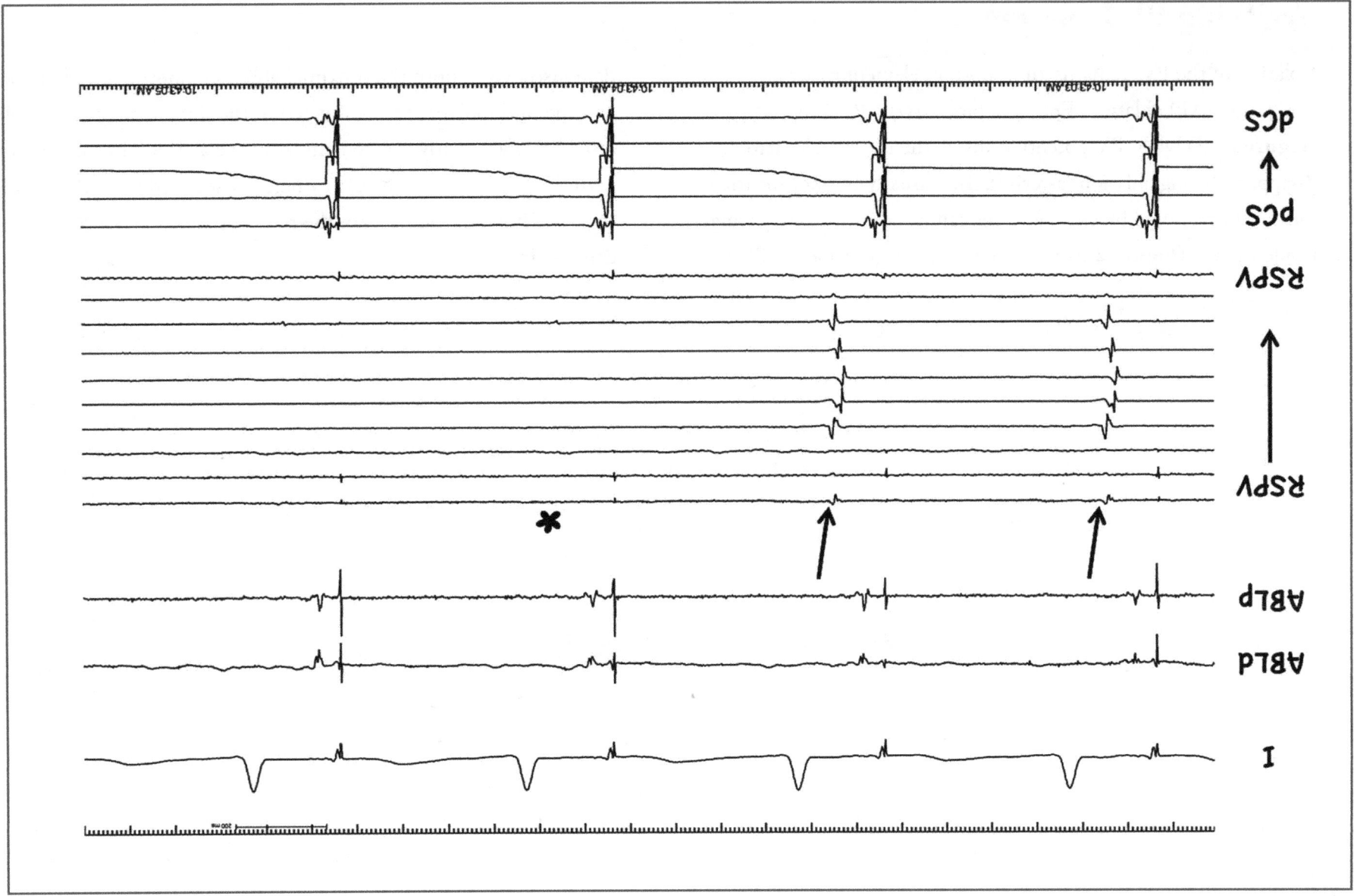

**Figure 2.23**

**Figure 2.24**

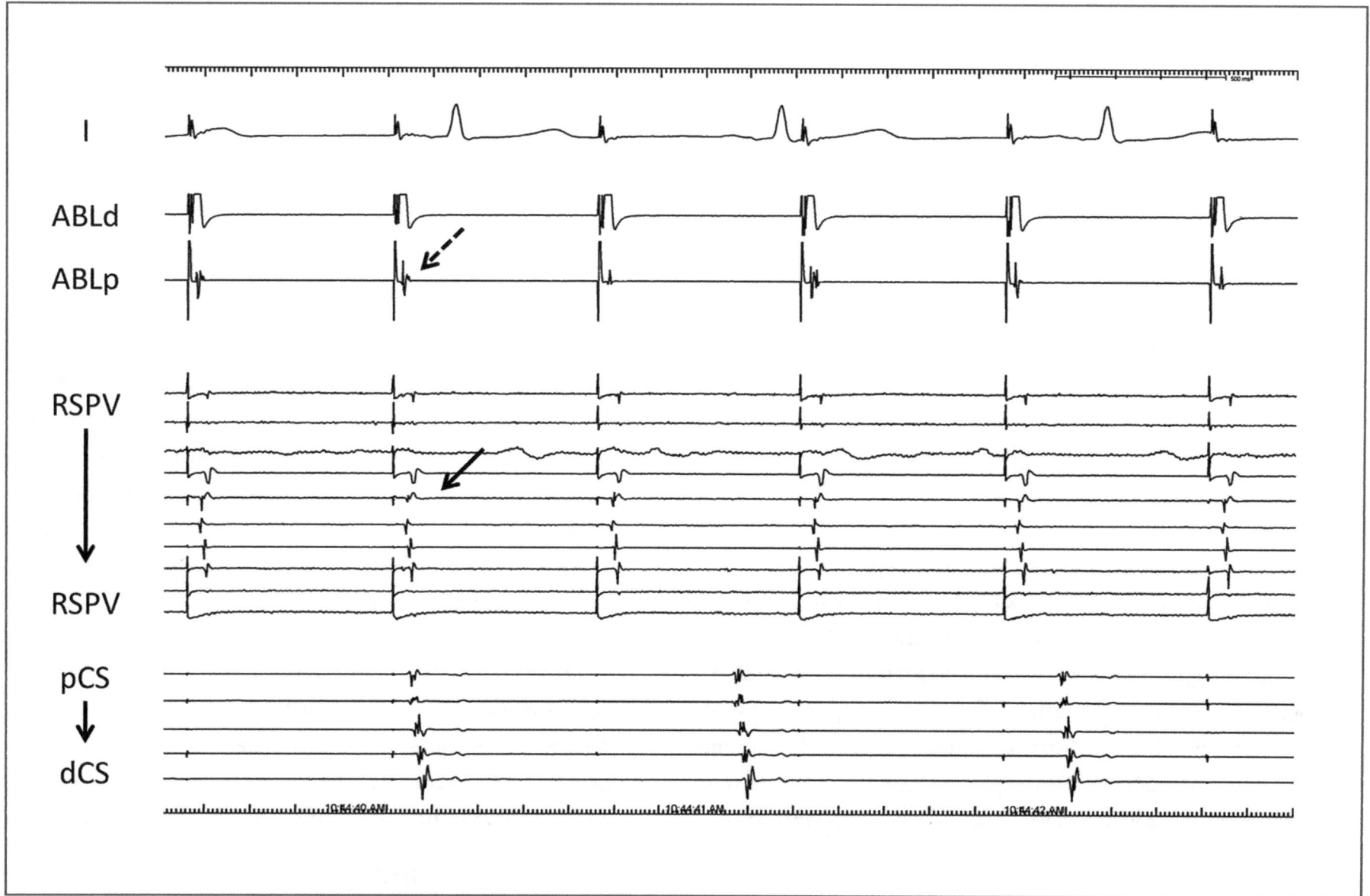

# Mapping Gaps in Conduction

Electrical reconnection of the PVs with the left atrium remains the most common finding in patients with AF recurrence following a PV isolation procedure. There may be single or multiple sites of electrical reconnection. Activation mapping of such gaps is an obvious step to achieve reisolation.

Typically, one of two strategies can be employed to find a gap in the circumferential line of block (**Figure 2.25**). One can pace the right or left atrium (Figure 2.25A) and determine the location of the earliest PV signals within the circumferential lesion set (1). Conversely, the PV can be paced (Figure 2.25B) and the earliest atrial signals outside the circumferential lesion set can be mapped (2).

In **Figure 2.26**, the region on the previous ablation line in the vicinity of the earliest PV potential (arrow) is mapped and the fractionated, prolonged activation (asterisk) very suggestive of a "gap" is found. Ablation at that site quickly results in a block to the PV (**Figure 2.27**) with the appearance of an isolated PV potential (asterisk) and apparent exit and entrance block. The remaining potentials following the stimulus artifact in the PV in the fourth cycle (arrow) are "far-field" atrial potentials from the LA.

**Figure 2.25**

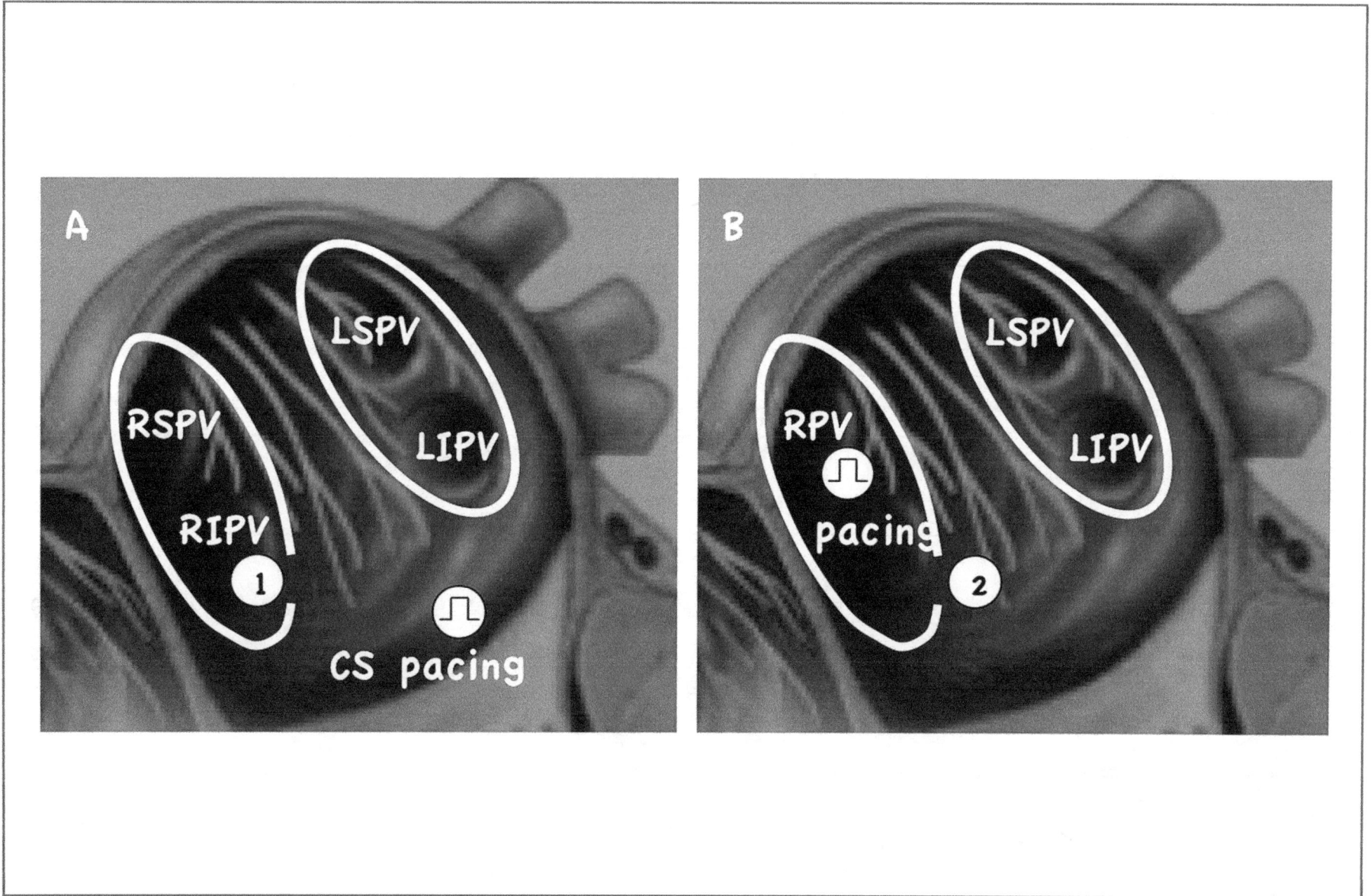

Figure 2.26

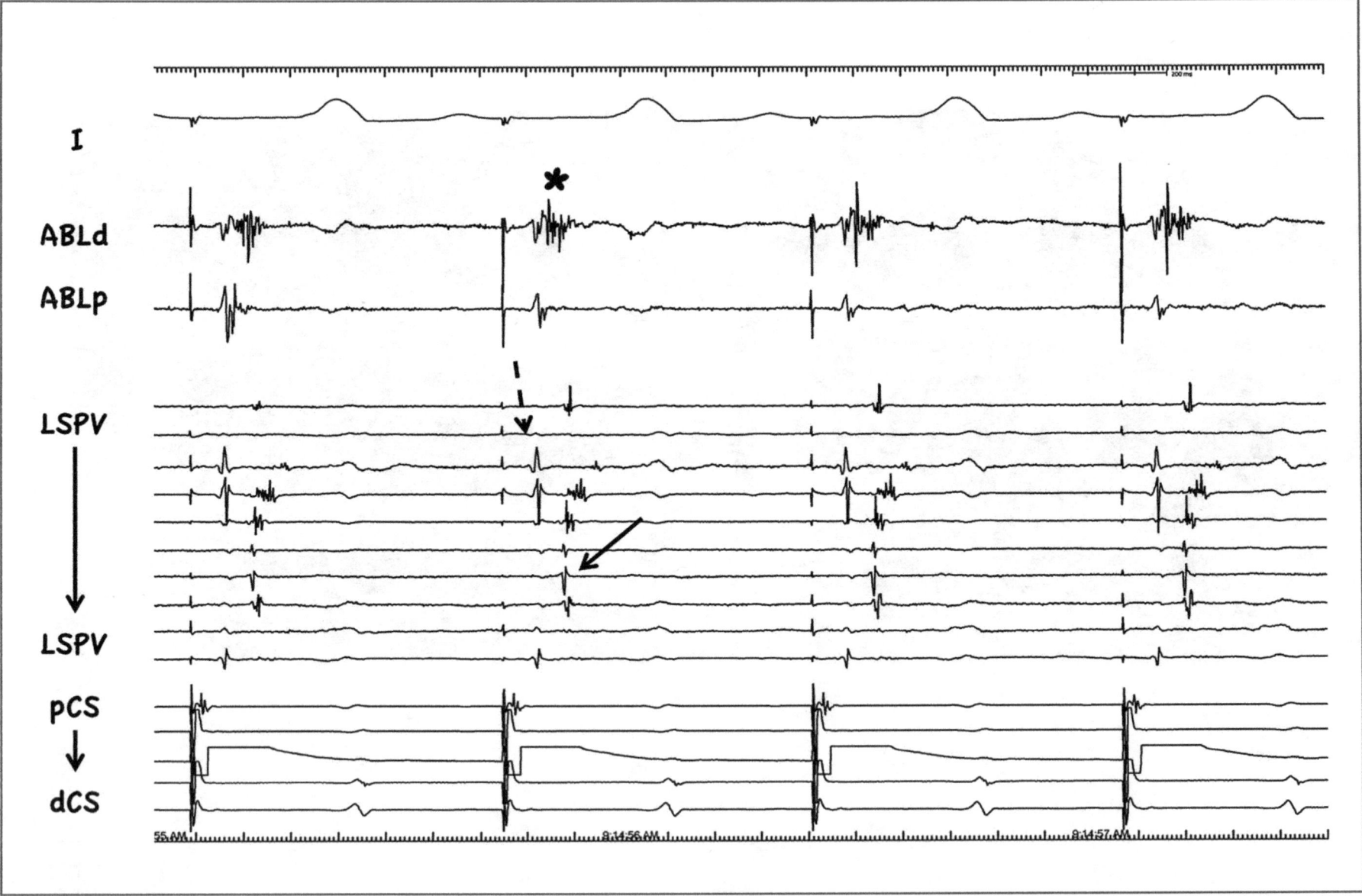

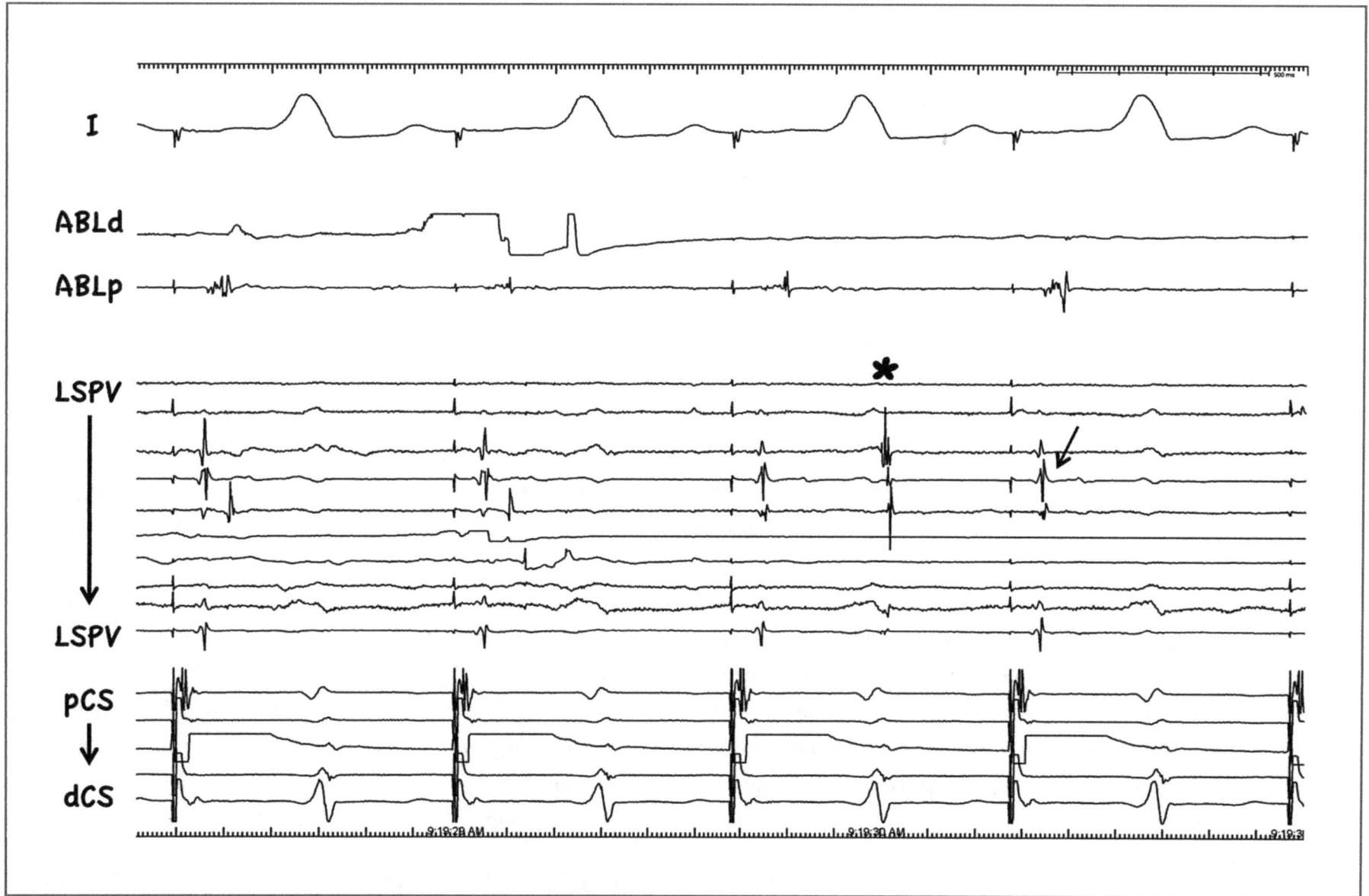

Figure 2.27

# Differential Pacing to Identify Near- vs. Far-Field EGMs

There are challenges confirming entrance or exit block due to lack of clear EGMs or misinterpretation of far-field EGMs. When trying to assess conduction block entering the PVs, one needs to distinguish true PV signals from EGMs caused by adjacent structures. For the left PVs, adjacent structures would be left atrium, LA appendage, or left ventricle. For the right PVs, apparent PV potentials may be caused by left atrium, right atrium, or the superior vena cava.

In general, local PV EGMs have a "near-field" appearance with a high d$V$/d$t$ ("sharp"), whereas adjacent structures have a "far-field" appearance with a lower d$V$/d$t$. Unfortunately, due to relative proximity of adjacent structures, EGM appearance alone may be unreliable and a differential pacing maneuver is required to distinguish the two.

*The preferred maneuver requires pacing the structure, which may be the source of the far-field EGM, and comparing this to the timing of the putative EGM to see if they match.* If the timing is the same, it is reasonable to assume that the EGM in question is nearer to the pacing site. Referring to **Figure 2.28**, if the mapping catheter is in the left superior PV (1), one might pace the LAA to see if the potential recorded in the PV moves in to match the timing of the paced site (and thus interpreted as coming from the LAA region) or not.

This is illustrated in **Figure 2.29**. Figure 2.29A shows CS pacing. The signals in the multipolar mapping catheter (dashed arrow) in the left superior PV may be PV or may be far-field LAA signals as suggested by their timing (solid arrow).

The LA appendage is paced (Figure 2.29B) and the LAA EGMs (solid arrow) are captured. However, the EGMs in question are markedly delayed (dashed arrows) related to that in LAA, suggesting they are NOT coming from LAA but rather from the PV.

**Figure 2.28**

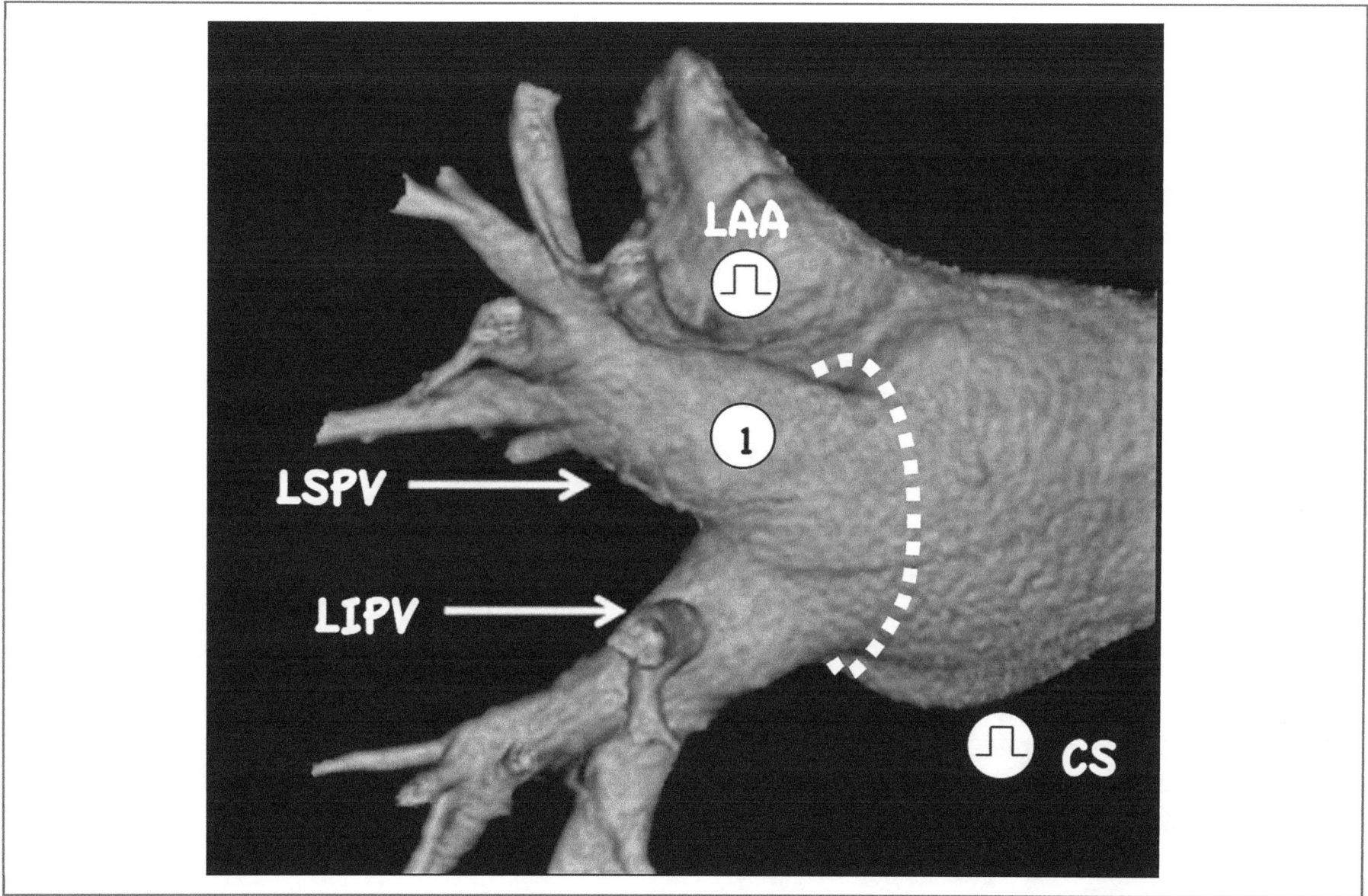

**Figure 2.29**

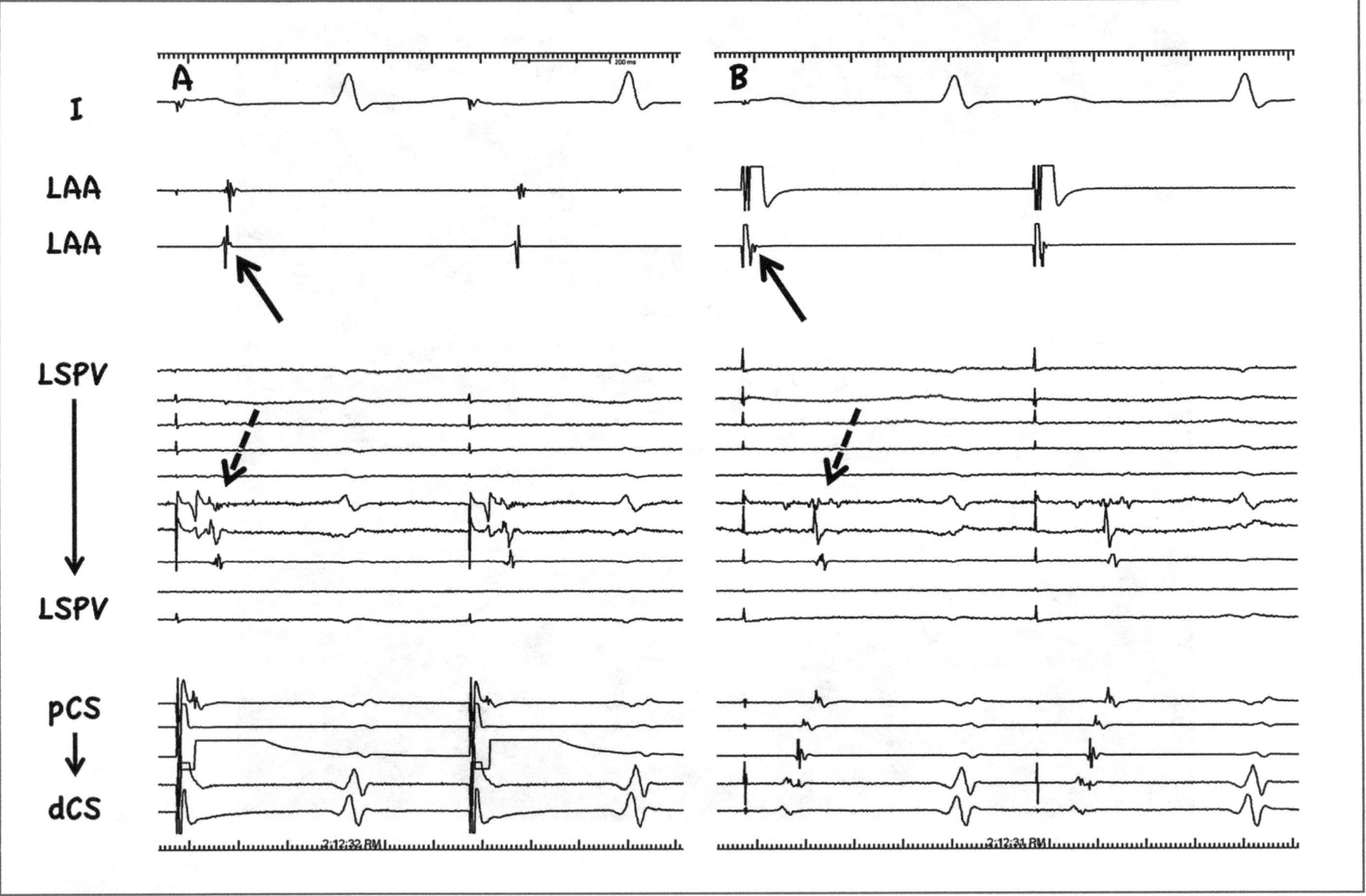

Conversely, **Figure 2.30A** once again shows CS pacing with signals in the mapping catheter (arrow) in the left superior PV of unclear origin. With LA appendage pacing, **Figure 2.30B**, the EGMs in question get "pulled in" (dashed arrow) to match the timing of the LAA pacing stimulus artifact. Hence, the signal in question was far-field LAA and no signals appear in the region of interest, confirming PV entrance block.

**Figure 2.31** is an example from the right PVs. Figure 2.31A shows CS pacing with EGMs in the mapping catheter (solid arrow) highly suspicious for true local PV potentials. However, pacing a potential far-field structure, the superior vena cava, gives rise to the EGMs in Figure 2.31B. During SVC pacing, the EGMs in question are "pulled in" to the stimulus artifact, clearly showing they are coming from that region with the PV now clear of any EGMs.

Clearly, unnecessary ablation can be avoided by identifying when the isolation has already been achieved.

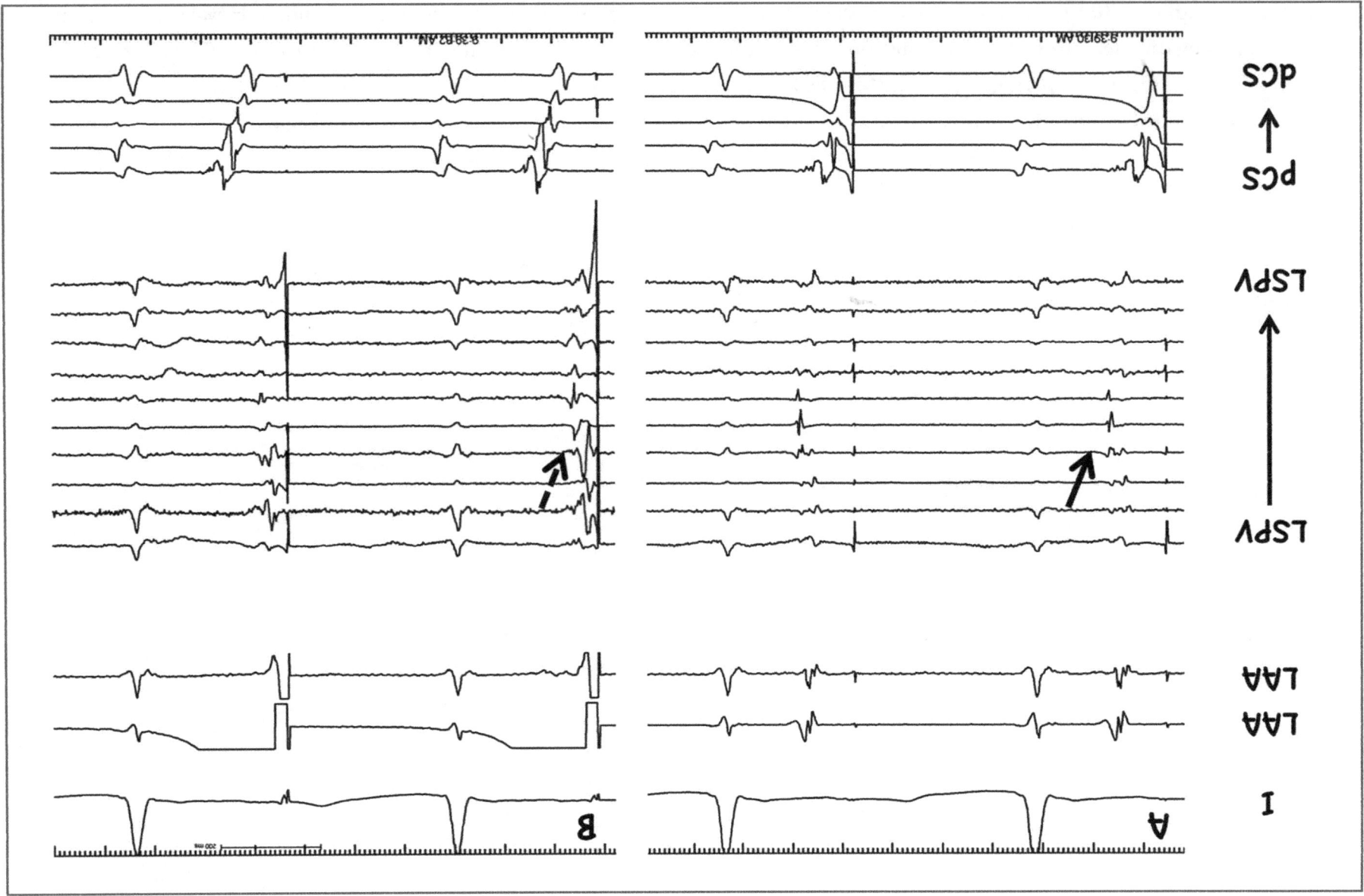

Figure 2.30

**Figure 2.31**

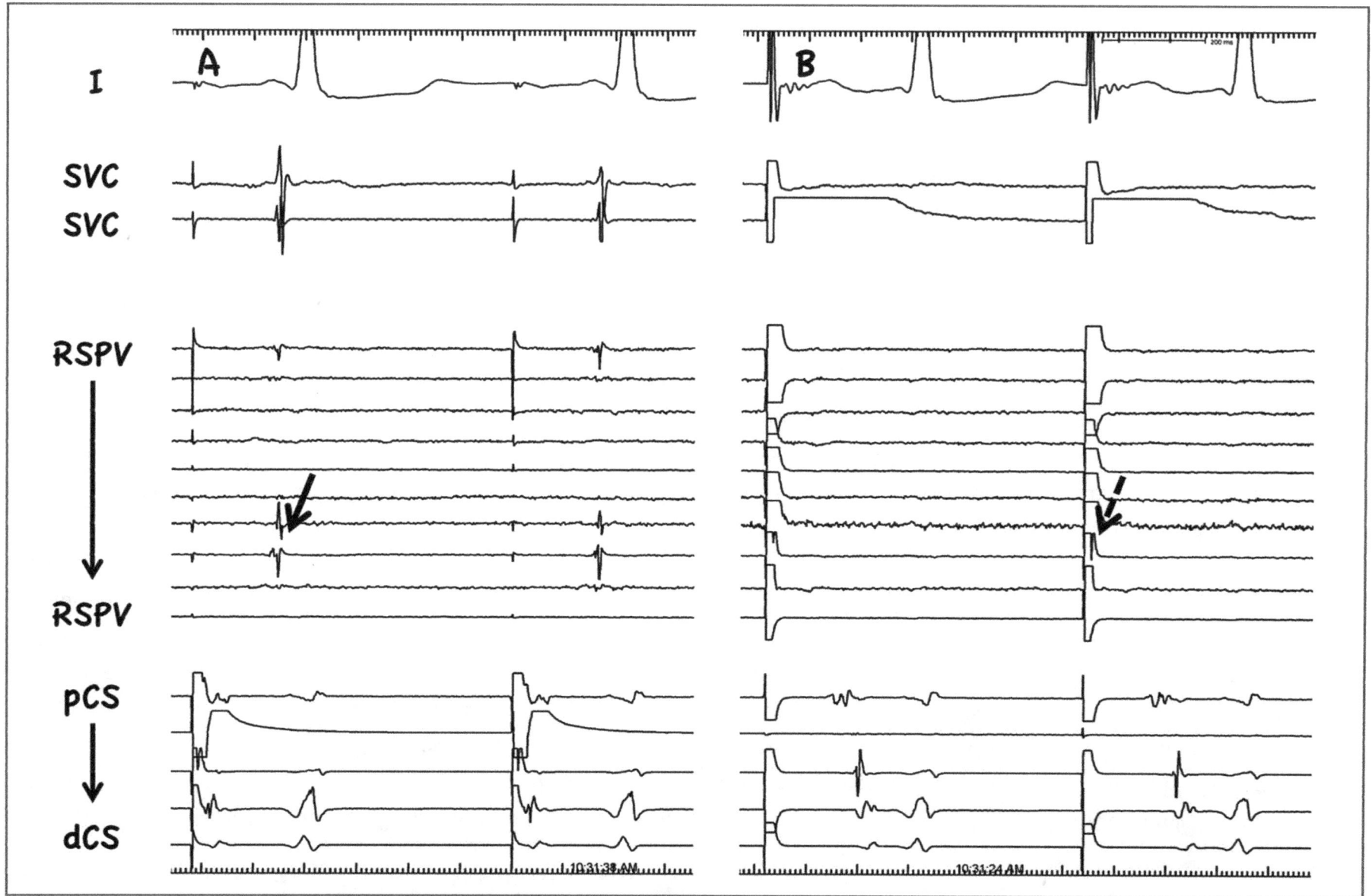

# Challenges Confirming Exit Block from the PV

Achieving pacing within the PV may be difficult due to inhomogeneity of muscle and complex architecture within the vein and the effects of ablation. In the example of **Figure 2.32**, the left superior PV is paced and the CS clearly demonstrates LA activity, which is dissociated from the pacing output. However, the presence of exit block depends on whether there is actually PV capture or not. It is possible that the artifact after the pacing stimulus is related to the high pacing output (inset) instead of local capture. When capture is not convincing or ambiguous, one can fall back on pacing at other sites within the PV or at least within the ablation line of the circumferential lesion. This was done and shown in **Figure 2.33**. In this case, the left inferior PV (also within the circumferential line) was paced while recording was continued in the LSPV. Now there is obvious capture of the LSPV evident due to the conduction time between the LIPV and LSPV. Note that capture at the actual pacing site itself in the LIPV is still ambiguous! It is now obvious that the left atrium (CS recordings) is dissociated from pacing within the lesion boundary.

Pacing from a site somewhat remote from the recording site but still within the boundary of the lesions will usually provide a clear indication of whether one is capturing or not.

**Figure 2.32**

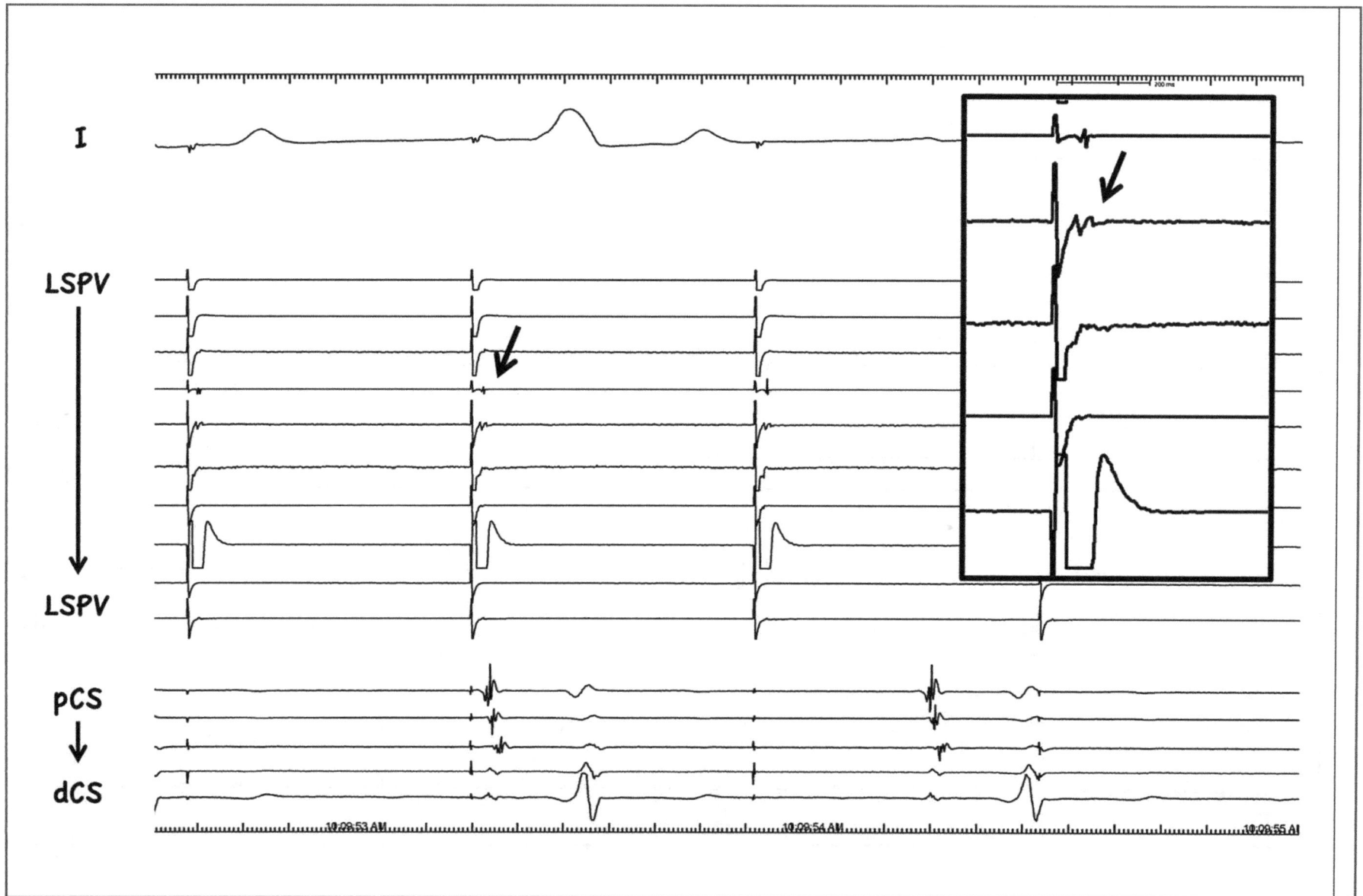

**Figure 2.33**

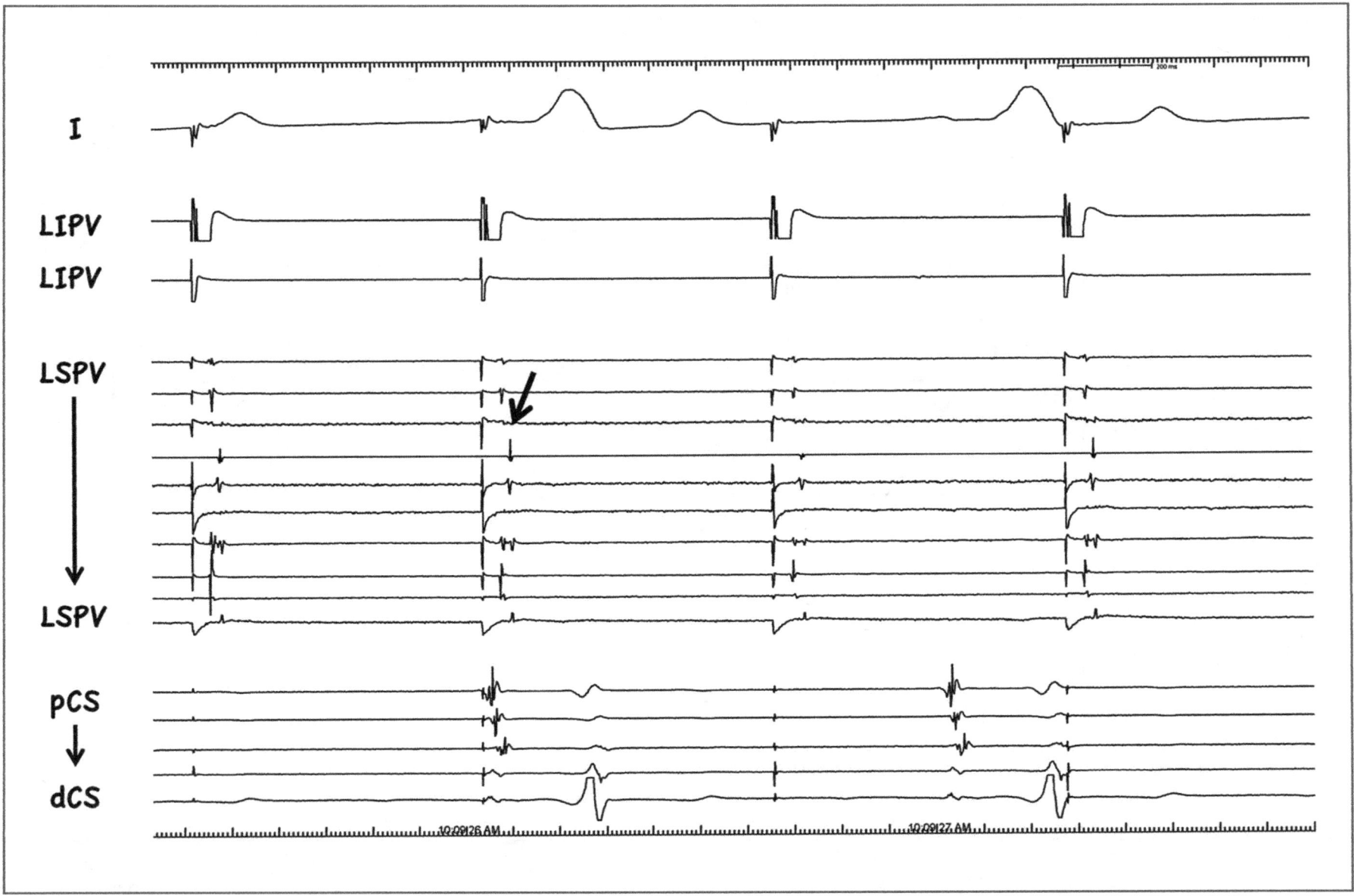

## Suggested Readings

1. Hirao K, Otomo K, Wang X, et al. Para-Hisian pacing: A new method for differentiating retrograde conduction over an accessory AV pathway from conduction over the AV node. *Circulation*. 1996;94:1027–1035.

2. Obeyesekere M, Leong-Sit P, Skanes A, et al. Determination of inadvertent atrial capture during para-Hisian pacing. *Circ Arrhythm Electrophysiol*. 2011;4(4):510–514.

3. Shah D, Haissaguerre M, Jais P, et al. Left atrial appendage activity masquerading as pulmonary vein potentials. *Circulation*. 2002;105(24):2821–2825.

4. Shah D, Burri H, Sunthorn H, Gentil-Baron P. Identifying far-field superior vena cava potentials within the right superior pulmonary vein. *Heart Rhythm*. 2006;3(8):898–902.

5. Shah D, Haïssaguerre M, Takahashi A, et al. Differential pacing for distinguishing block from persistent conduction through an ablation line. *Circulation*. 2000;102(13):1517–1522.

6. Obeyesekere M, Gula LJ, Modi S, et al. Tachycardia induction with ventricular extrastimuli differentiates atypical atrioventricular nodal reentrant tachycardia from orthodromic reciprocating tachycardia. *Heart Rhythm*. 2012;9:335–341.

7. Bennett MT, Gula LJ, Klein GJ, et al. An alternative method of assessing bidirectional block for atrial flutter. *J Cardiovasc Electrophysiol*. 2011;22(4):431–435.

8. O'Neill M, Veenhuyzen G, Knecht S. *Catheter Ablation of Persistent Atrial Fibrillation – A Practical Guide*. London UK: Remedica Publishing; 2008.

# 3

# Extrastimuli during Tachycardia

Programming atrial or ventricular extrastimuli into a tachycardia has been a "standard" maneuver for many years and one of the most useful. As with entrainment, ventricular extrastimuli will generally be more useful than supraventricular extrastimuli. An extrastimulus might be thought of as "overdrive pacing" for one cycle that may entrain for one cycle, or alternatively, just overdrive without entrainment. An advantage over entrainment testing lies in the ability to scan the whole tachycardia cycle at multiple coupling intervals relatively rapidly, although fundamentally extrastimuli provide the same information as entrainment, as emphasized in chapter 1. Ectopic beats may perturb other elements to reveal tachycardia mechanism, as will be illustrated. One might expand the use of the principles learned from extrastimuli to observe the effects of spontaneous ectopics since fundamentally the same elements exist, thus allowing one to interpret the ECG more "physiologically."

## Relationship of Extrastimuli to Entrainment

**Figure 3.1** shows a supraventricular tachycardia (SVT) with right bundle branch block (RBBB) morphology and normal HV interval. The atrial activation sequence is eccentric (CS 5–6 earliest). A premature ventricular contraction (PVC) programmed into the cardiac cycle advances the subsequent atrial activation with the same atrial sequence. It is noted that the His deflection (short arrow) has not changed its timing or morphology, even though it follows the stimulus artifact proving that it was activated anterogradely and therefore "refractory" to retrograde activation by the PVC. This is the classical "His refractory" PVC that can only advance the subsequent atrial activation via an accessory pathway (AP), in this case, a left lateral AP. Theoretically speaking, this proves only that the AP is "present" but not participating in the tachycardia circuit since it may be advancing a focal atrial tachycardia (AT) near the AP or possibly a junctional tachycardia. From a practical viewpoint, the latter would be a rare occurrence indeed.

It is noted that the SA and VA intervals during tachycardia as well as the postpacing interval (PPI) from the right ventricular apex (RVA) are measured. Although the mechanism of this tachycardia is obvious, this is done to remind that this tachycardia has been essentially "entrained" for one beat and the intervals validated for entrainment apply. In this case, SA-VA interval is 95 and the PPI–tachycardia cycle length (TCL) is 150. These numbers would be greater than the usual published limit for a septal AP, hence indicating (in the absence of significant decrement in the pathway) that the right ventricular (RV) pacing site is "out" of the circuit. One might also expect SA-VA > 85 and PPI-TCL > 115 with atrioventricular nodal reentrant tachycardia (AVNRT) or atrioventricular reentrant tachycardia (AVRT) over a left AP (the latter's essential circuit of course including the anterograde left bundle branch (LBB) and the left AP).

## Figure 3.1

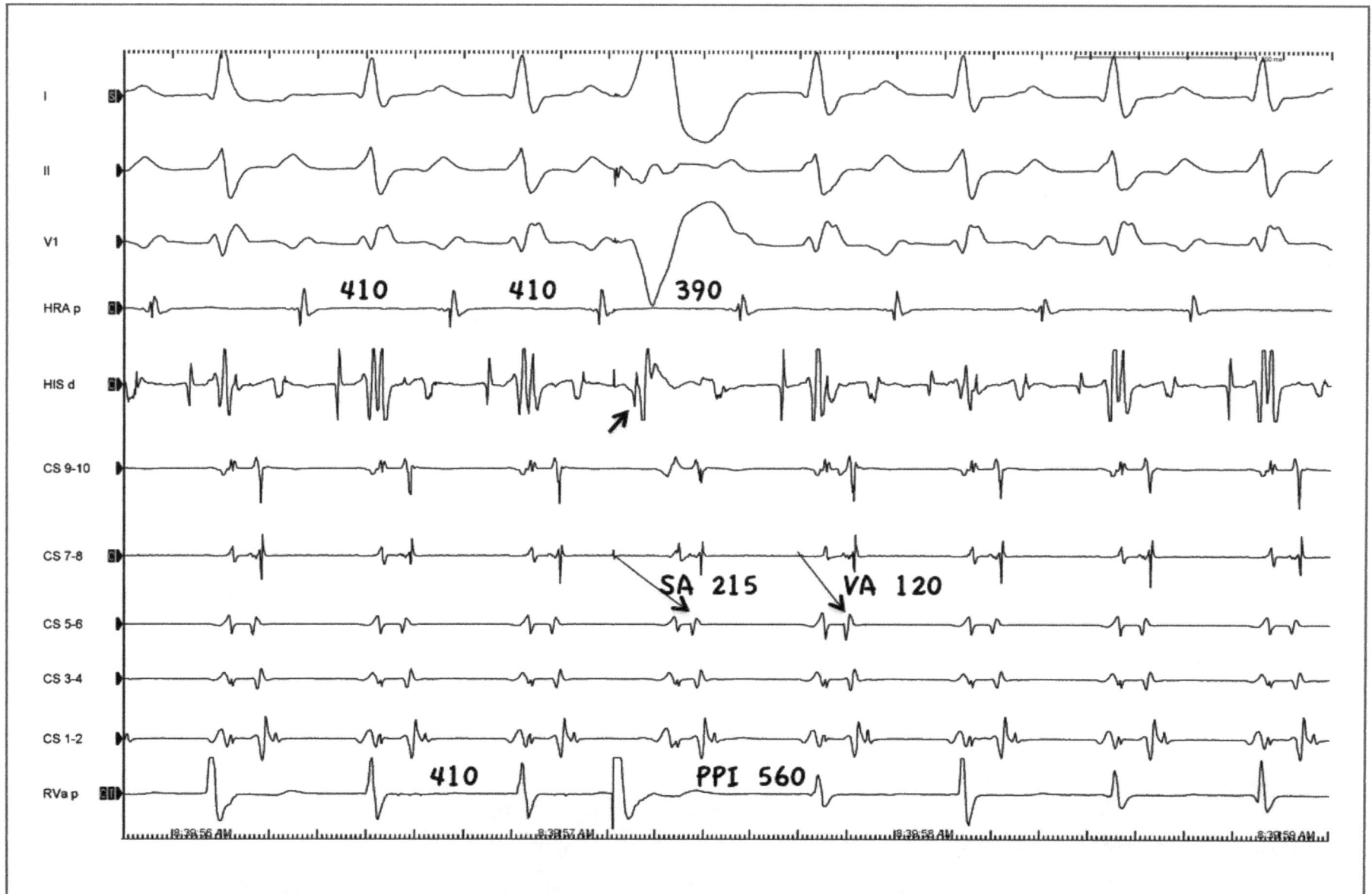

**Figure 3.2** shows a PVC inserted into a tachycardia again with a normal HV interval and eccentric retrograde atrial activation (CS 1,2 first). The first A after the PVC has been advanced (reset). In this instance, the His after the PVC has been advanced and the His morphology changed, indicating retrograde activation of the His bundle. *Thus, it can't be said that the reset A occurred during His refractoriness.* However, there are other indications that this has to be orthodromic atrioventricular (AV) reentry over a left lateral, including the retrograde atrial activation sequence after the PVC, which would not be expected with retrograde conduction over the normal His bundle and AVN. In addition, the difference between the PPI and TCL (70) and the SA-VA (also 70) suggest the RV

pacing site is relatively "in" the circuit in this instance and not compatible with AVNRT. This is a logical extrapolation of the data presented in the original paper of Michaud et al showing that all AVNRT patients in their series had values for these intervals above 85 ms (SA-VA) and 115 ms (PPI-TCL).

One might ask how the atrium was reset over the left AP even though the His bundle was activated retrogradely by the PVC. In this case, one might guess that the antidromic wave provided by the PVC blocked retrogradely in the AVN *and this region must have recovered excitability when the next atrial wave came around.*

**Figure 3.2**

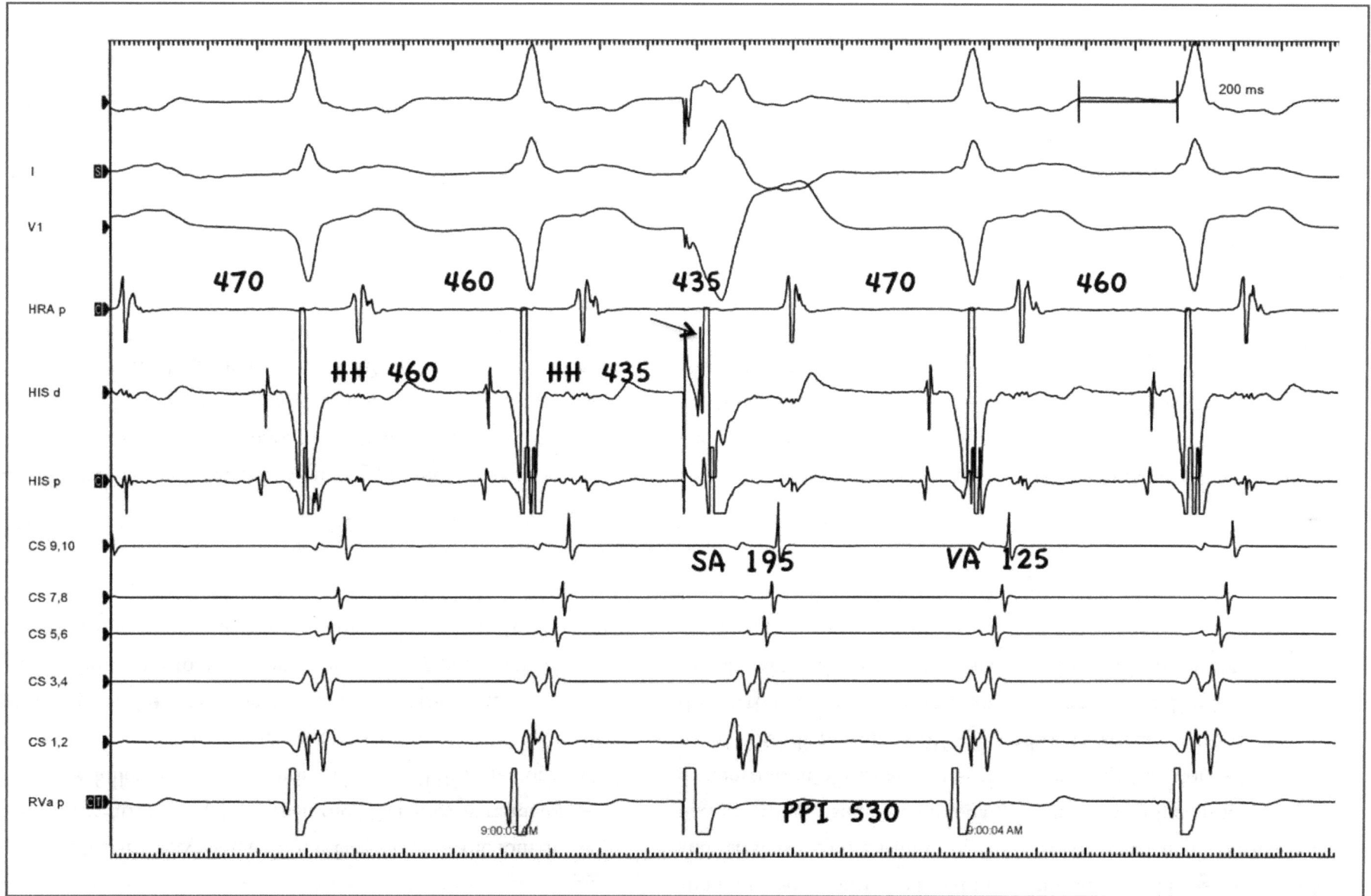

**Figure 3.3** shows SVT with central atrial activation sequence. The first 3 stimulus artifacts shown do not conduct, but the fourth captures the ventricle at a time when the His is clearly refractory and advances the following A. Thus, it is a classical "His refractory PVC" advancing the next A and proving the presence of a septal AP in this instance. **Figure 3.4** shows the continuation of the sequence in Figure 3.3 showing it to be the onset of overdrive pacing, which resets with fusion (the first QRS is clearly different than fully paced

and moreover, "must" be fused by definition if the His is anterograde and refractory to the stimulus). The onset of ventricular overdrive pacing has been suggested to be useful in this way (i.e., reset with demonstration of fusion for less than 2 or so cycles prior to fully paced QRS diagnostic of AVRT), and this just makes it clear that this method is essentially a variant to programming PVCs into the cardiac cycle (which might in fact be simpler) and certainly works on the same principle.

## Diagnostic Utility of Programmed PVC, PAC: Specific Examples

**Figure 3.5** was shown in chapter 1 but is reproduced here to complete the spectrum of response to the PVC. A narrow QRS tachycardia with a long VA time and eccentric atrial activation (both the CS9, 10, and the high right atrium (HRA) are earlier than the His A) is shown. A His-synchronous PVC *delays* the next A and

resets the tachycardia. This unequivocally documents the presence and participation of the pathway in the tachycardia. This response can be seen in slowly conducting, decremental pathways. This pathway was successfully ablated at 7 o'clock on the tricuspid annulus.

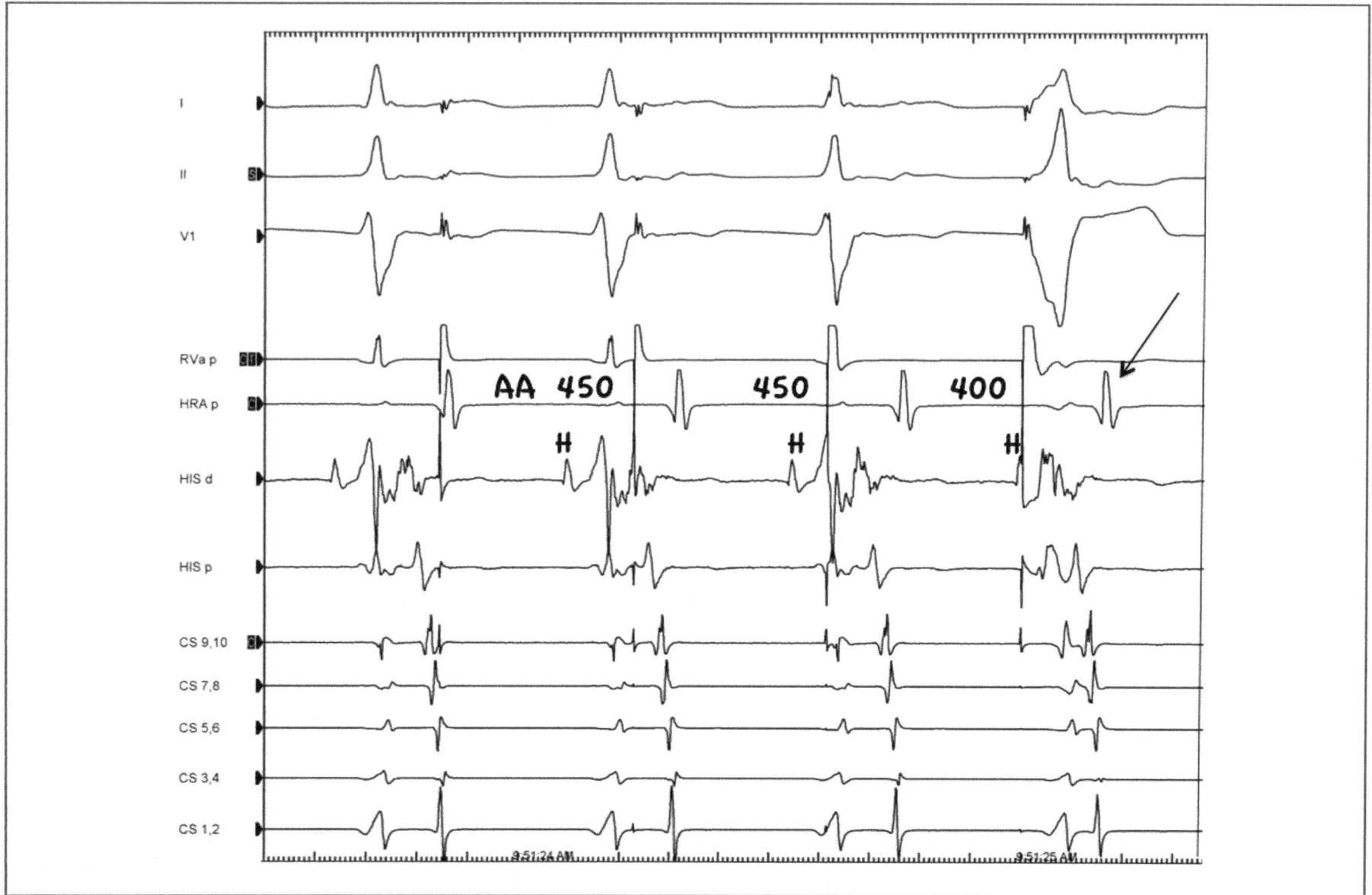

Figure 3.3

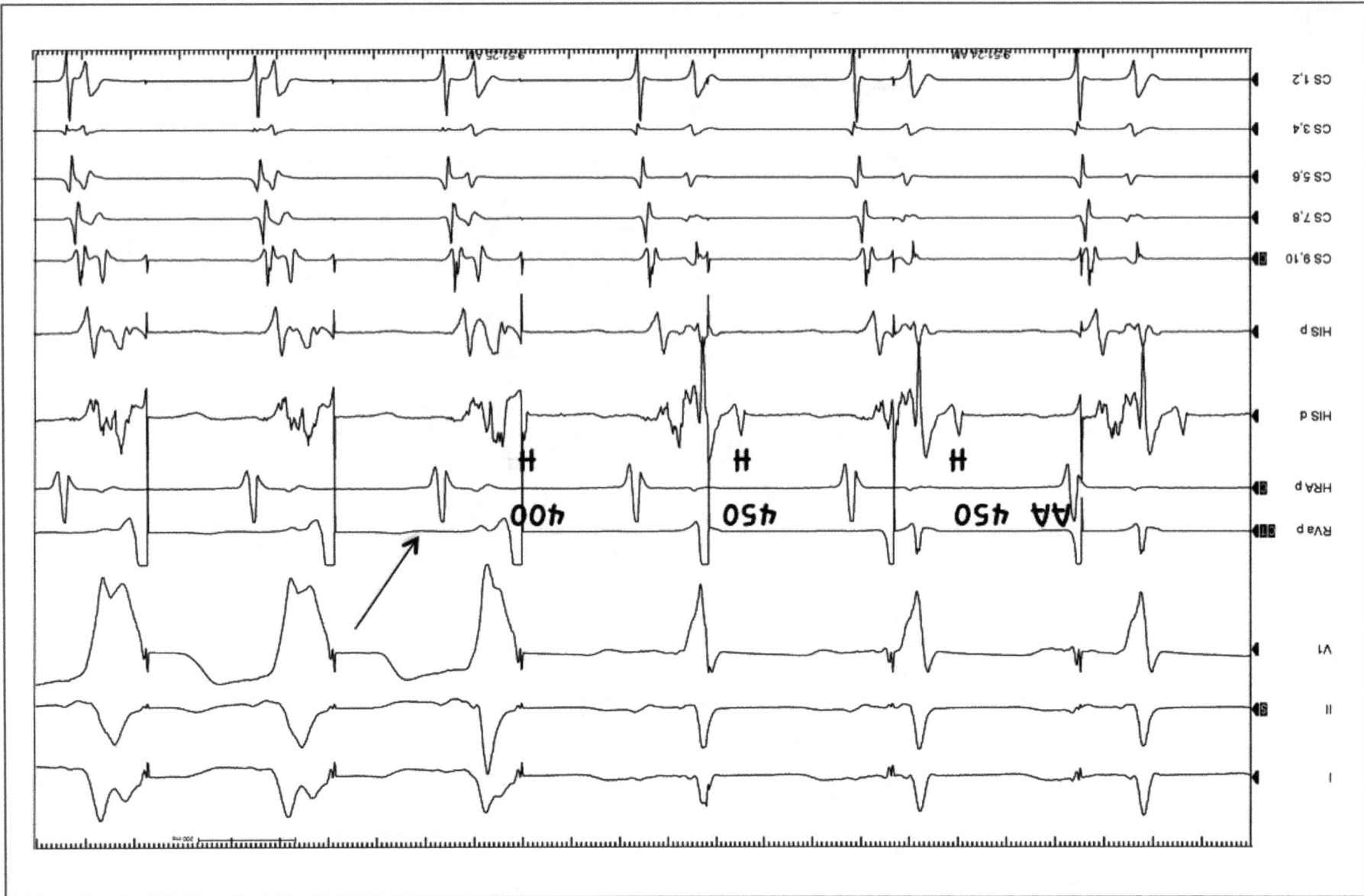

Figure 3.4

## Figure 3.5

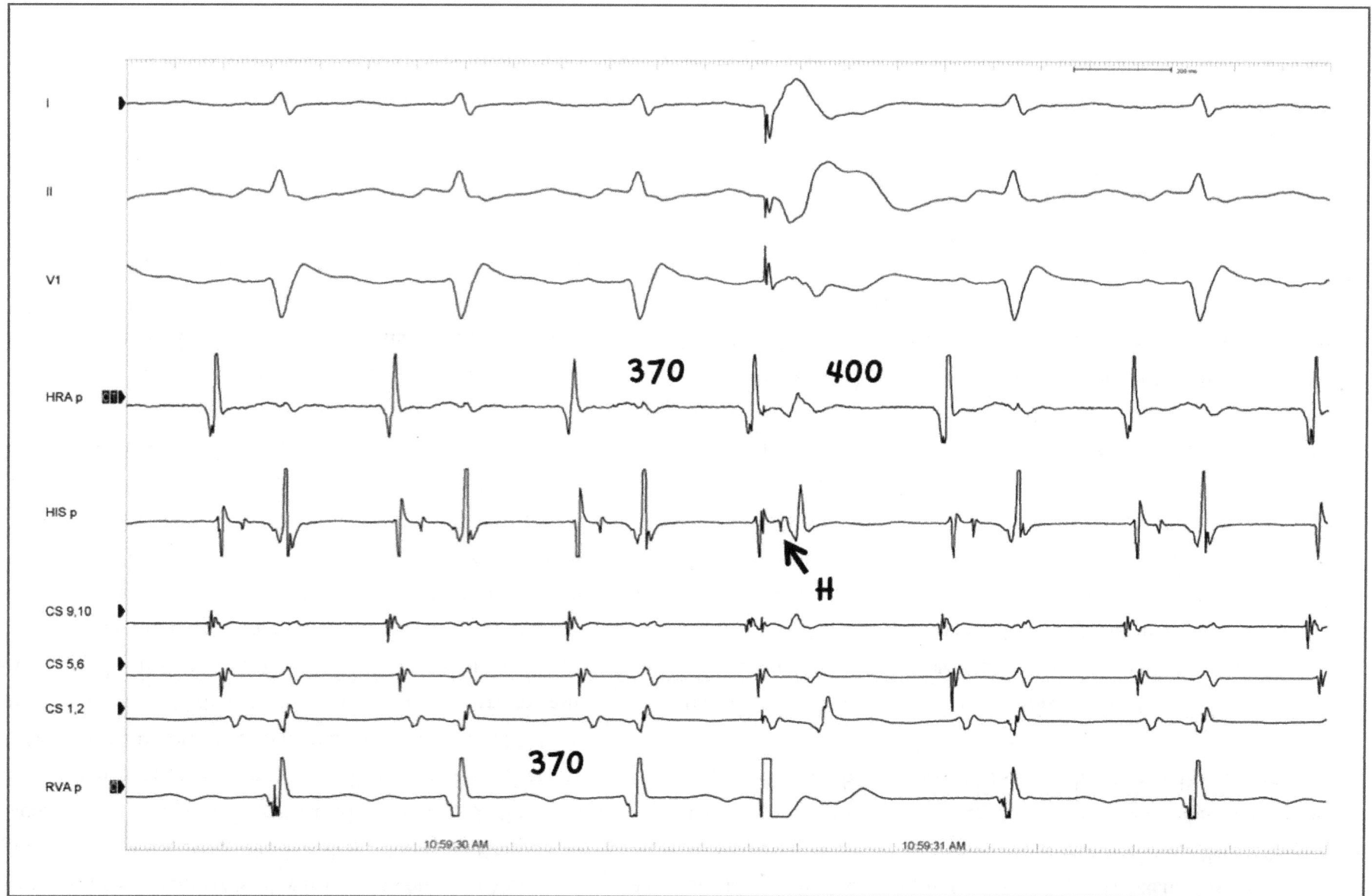

**Figure 3.6** shows a regular SVT with a late-coupled PVC programmed into the cycle. The His deflection is on time (refractory) and unaffected by the paced QRS. The QRS following the PVC is narrow and obviously "fused" and the His refractory. This PVC results in termination of the tachycardia. An SVT that terminates after a PVC that is not conducted to the atrium cannot be an AT. In this instance, termination indicates that the PVC could not have conducted retrogradely over the normal AV conducting system but reached a critical component of the circuit, i.e., an AP, and found it refractory thus terminating tachycardia. The latter was an obligatory limb of the circuit (i.e., must participate in the circuit). The usual proviso is that the observation must be reproducible as it was in this case.

**Figure 3.7** shows 2 consecutive late coupled PVCs programmed into a tachycardia circuit. The first may be "His refractory" and the second is clearly fused between the normal QRS and the wider PVC. The second PVC appears to delay the next A. However, a closer look reveals that the cycle length (CL) is widely variable in this tachycardia and no useful diagnostic information is available. Any change that occurs should be reproducible and believably beyond the range of the variability in cycle length of the observed tachycardia.

The deductive principles behind interpreting SVT with PVCs programmed into the cardiac cycle can be used when one sees spontaneous PVCs as in **Figure 3.8**. In this instance, an SVT with short VA interval is interrupted by a PVC. The PVC results in a VAV response, ruling out AT. The PVC is too early to be considered His refractory. Nonetheless, the very long (PPI minus TCL) and (VA paced minus VA tachycardia) suggest that the RV (the PVC has left bundle branch block (LBBB) pattern and left axis) is not part of the circuit and favors a diagnosis of AVNRT (see chapter 1). Of course, caution is warranted with such a closely coupled PVC since it may introduce "decrement" into the circuit and prolong the intervals even in AVRT. In this case, it was AVNRT.

**Figure 3.6**

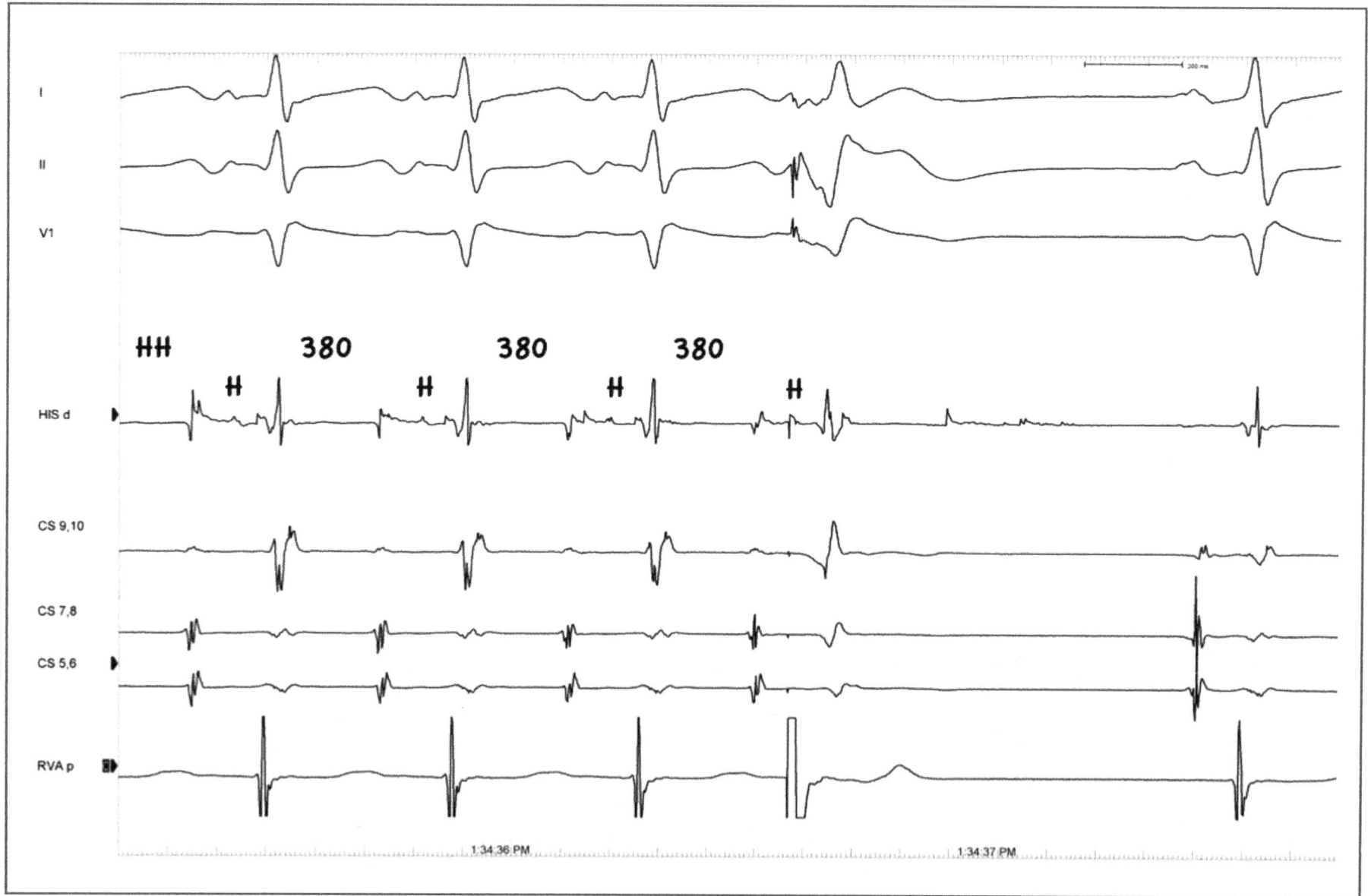

**Figure 3.7**

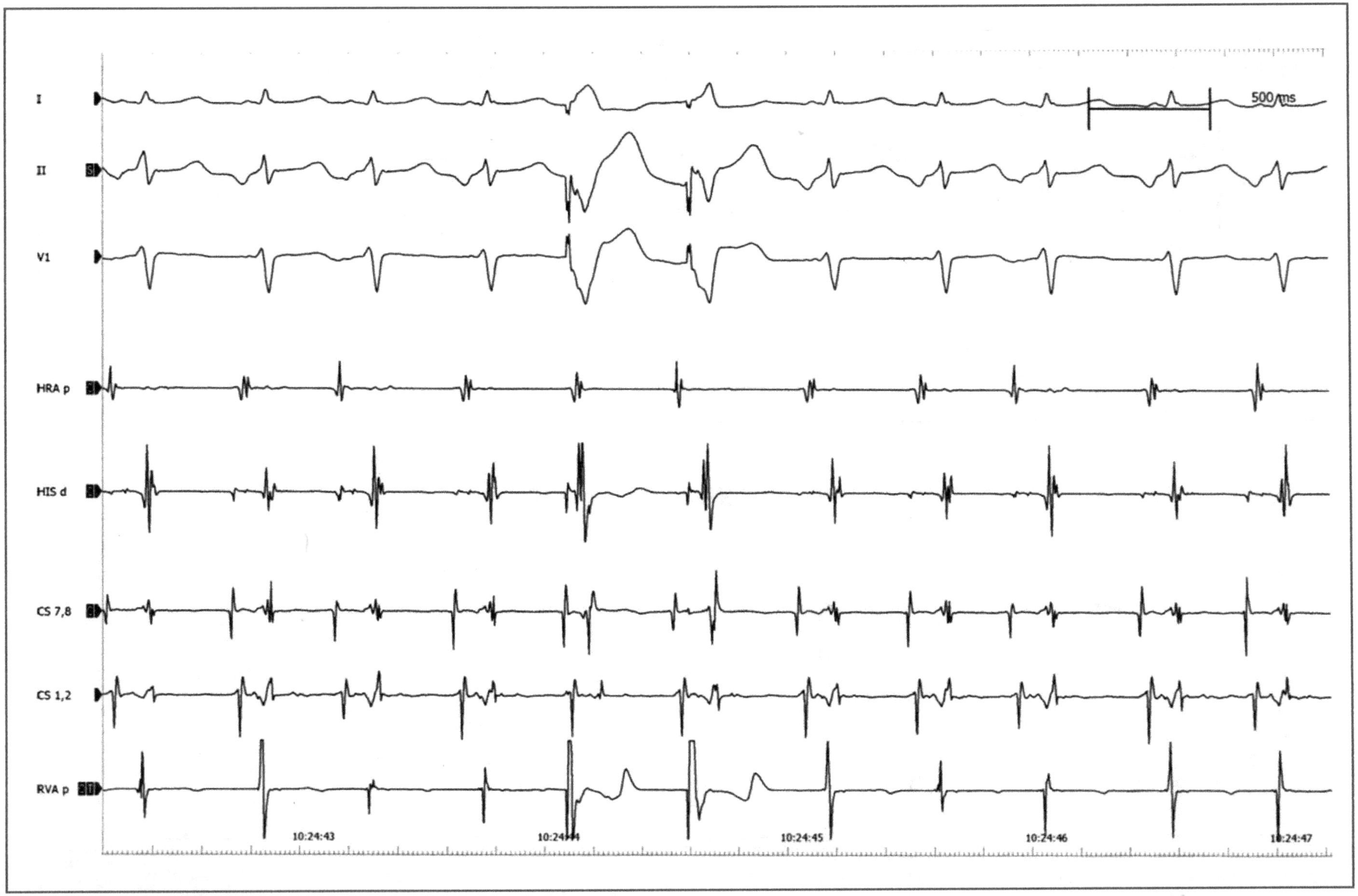

## Figure 3.8

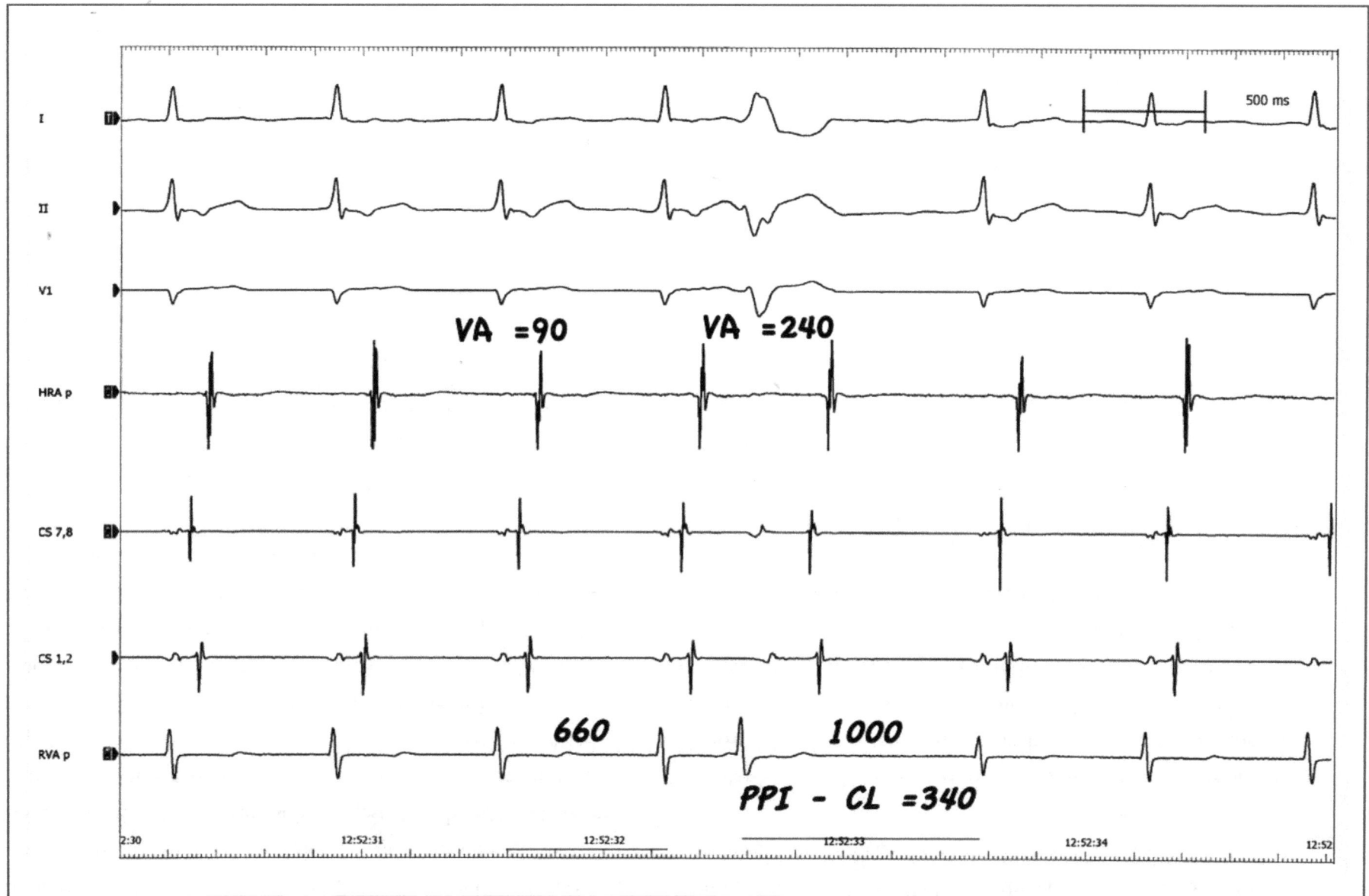

The remarkable observation in **Figure 3.9** is that the PVC is very late coupled, barely affecting the QRS on the surface ECG, yet advancing the next A. This can only suggest that the RV PVC is very close to the excitable gap of the circuit, a short distance from the retrogradely conducting AP, in this case a right AV pathway. A left AV pathway under similar circumstances would require that the PVC be more premature to reset the tachycardia since it obviously has a longer distance to the AP. In fact, it may not be possible to preexcite some left-sided pathways from the RV during AVRT at the time when the His is refractory for this reason!

The His-synchronous PVC in **Figure 3.10** apparently terminates tachycardia. Closer inspection shows that the PVC falls too close to the timing of the next destined A (dots) to have influenced it. The termination of tachycardia here is therefore fortuitous and no conclusion can be made on the influence of the PVC.

Regular SVT with earliest atrial activation in the HRA is present in **Figure 3.11**. The differential diagnosis is AVRT using a right-sided AP or AT. There is minor cycle length variability but a very early-coupled PVC from the RVA fails to impact the timing of the atrial activation. One would expect that such an early RV PVC would have excellent access to the circuit with a right AP to reset it but it did nothing. Although it is possible that conduction delay in the VA time *exactly* offsets the prematurity of the ventricular extrastimulus, it is more probable that it merely dissociated the V from an AT, which was the case here.

**Figure 3.9**

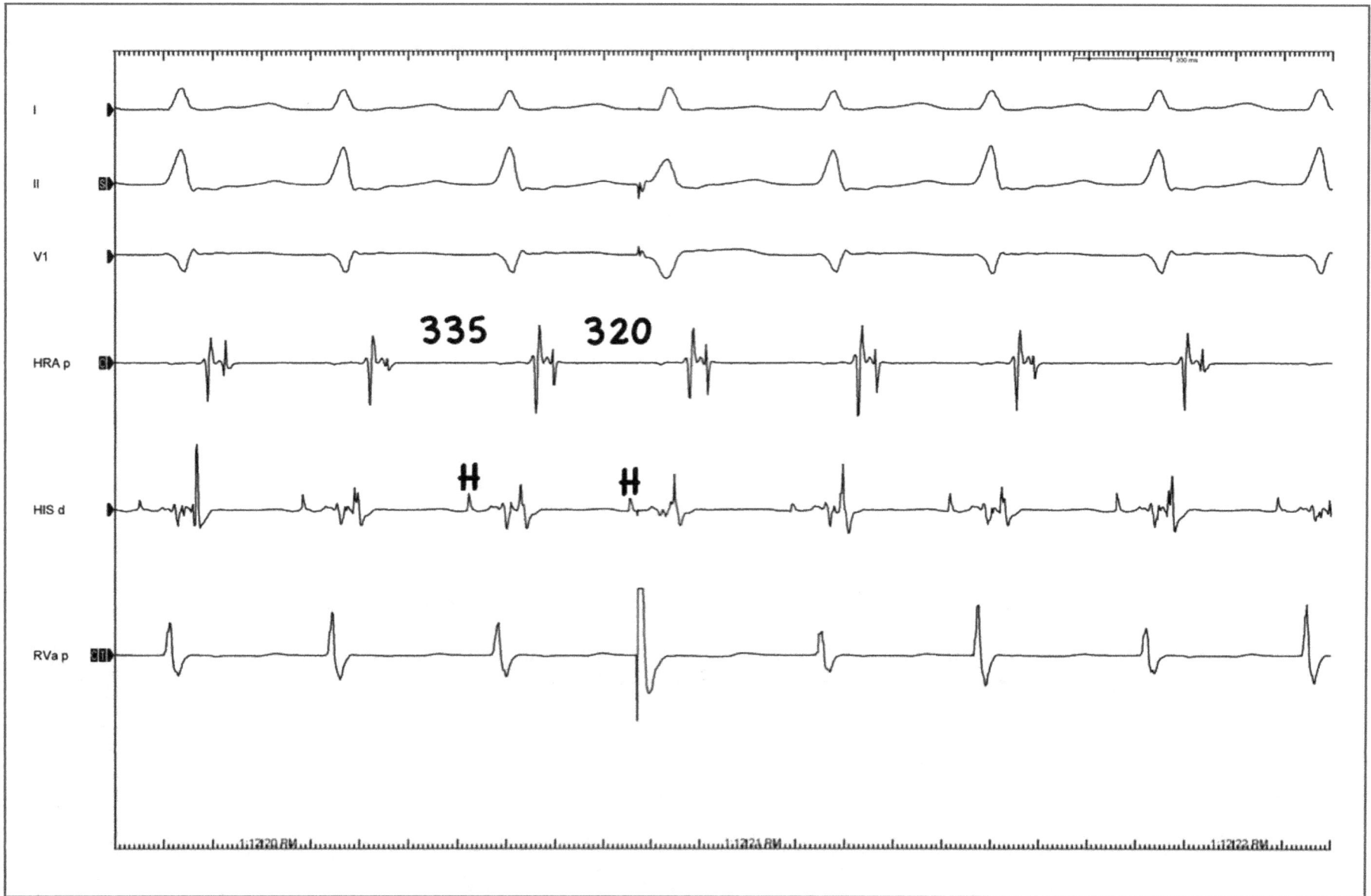

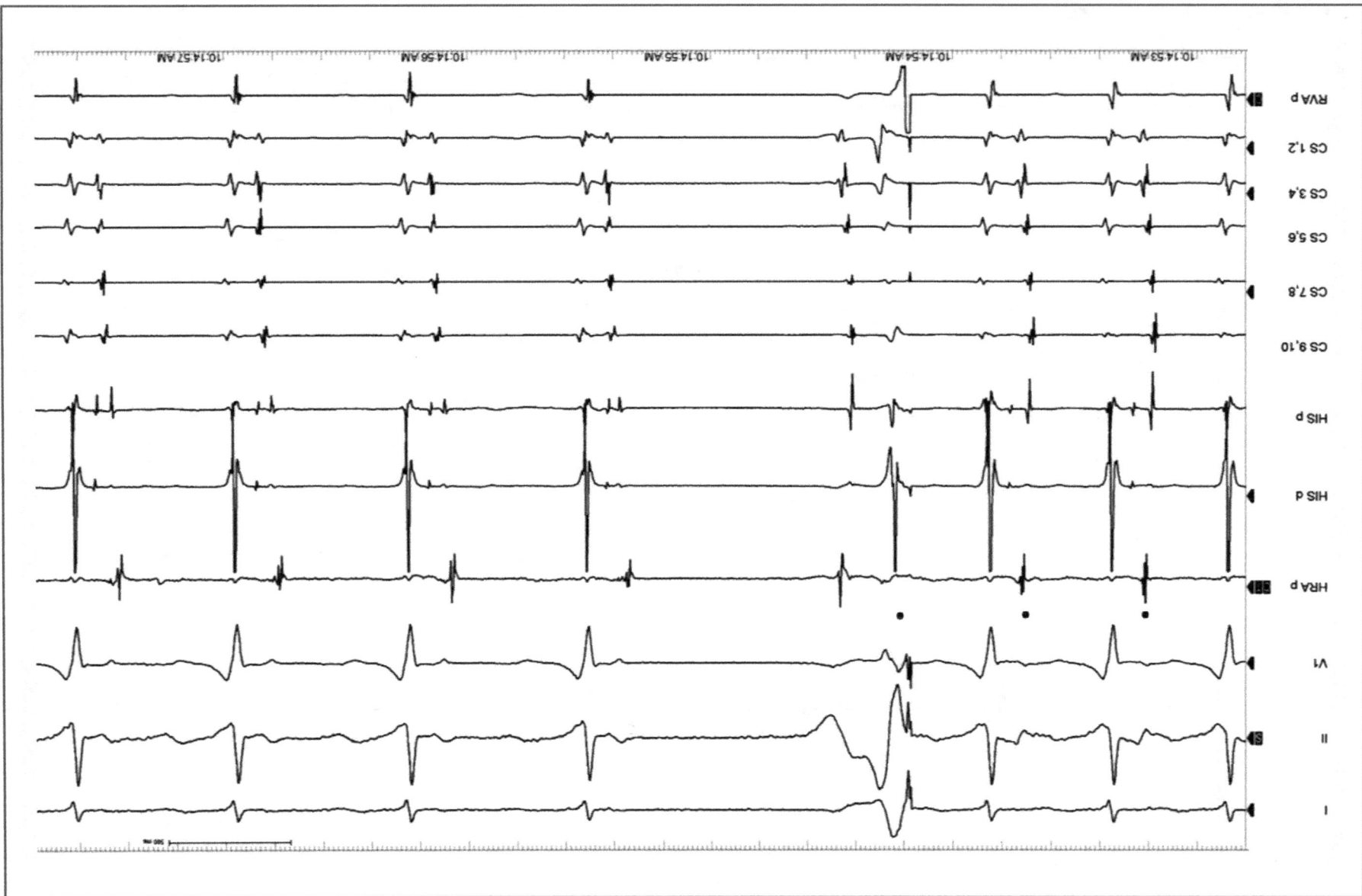

**Figure 3.10**

Figure 3.11

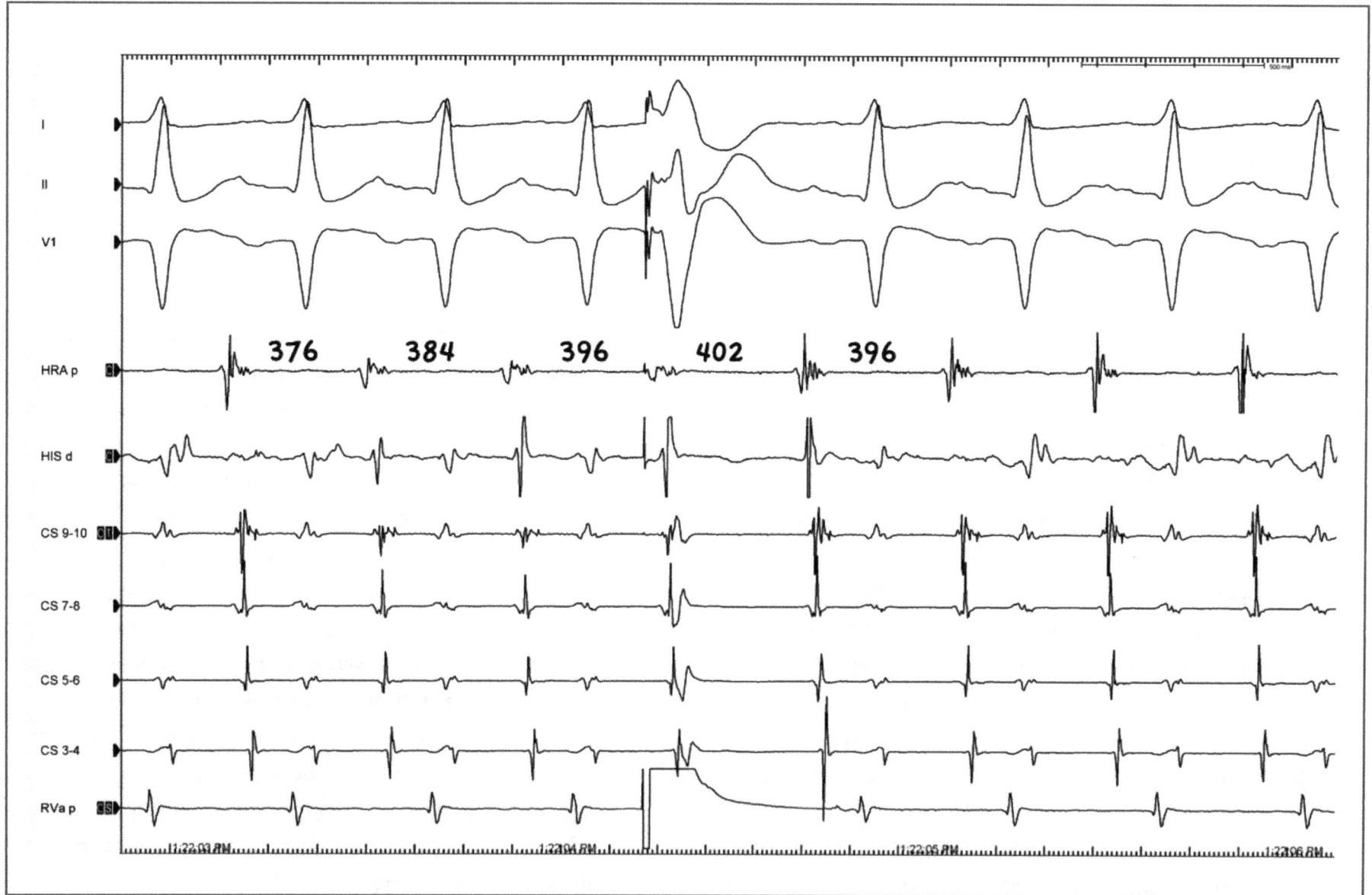

The coronary sinus (CS) electrograms (EGMs) are not necessary to make the diagnosis in **Figure 3.12**. The beginning of the tracing shows SVT with LBBB and normal AH interval. A PVC timed when the His is refractory readily preexcites the next A, confirming presence of an AP. The QRS then becomes normal and the tachycardia continues with a shorter cycle length. The VA interval is concomitantly shorter (225 to 170 ms measured at the His site), confirming the participation of the LBB in the tachycardia circuit. The shortened VA is possible due to the shortened circuit length when LBB resolves (**Figure 3.13**). The cycle length shortens in this scenario but may actually not change or lengthen depending on the concomitant AH change when QRS normalizes. It is reasonable to hypothesize that the RV PVC preexcited the LB via transseptal conduction, allowing it more time to recover excitability when the next supraventricular beat arrived ("peeling back"). LBBB during AVRT over a left AP is readily preexcited from the RV (i.e., with relatively late coupled PVC) since the RV and RBB become an integral part of the circuit in this situation.

The tachycardia in **Figure 3.14** shows a central atrial sequence with the His bundle (HB) atrial EGM first. The PVC may not be His refractory (not seen on this cycle) but nonetheless advances the A. This results in marked prolongation of the AH interval but with continuation of the tachycardia. The interesting observation is that the first cycle after the long AH shows a change in atrial activation exposing a left lateral AP. This is only present for one cycle, presumably because the left AP had a longer refractory period and subsequent shortening of the AH on the next cycle resulted in failure of retrograde activation over the left AP. This patient did have dual AV node pathways and a left and septal AP, and the left pathway was brought to light by the former observation.

Figure 3.12

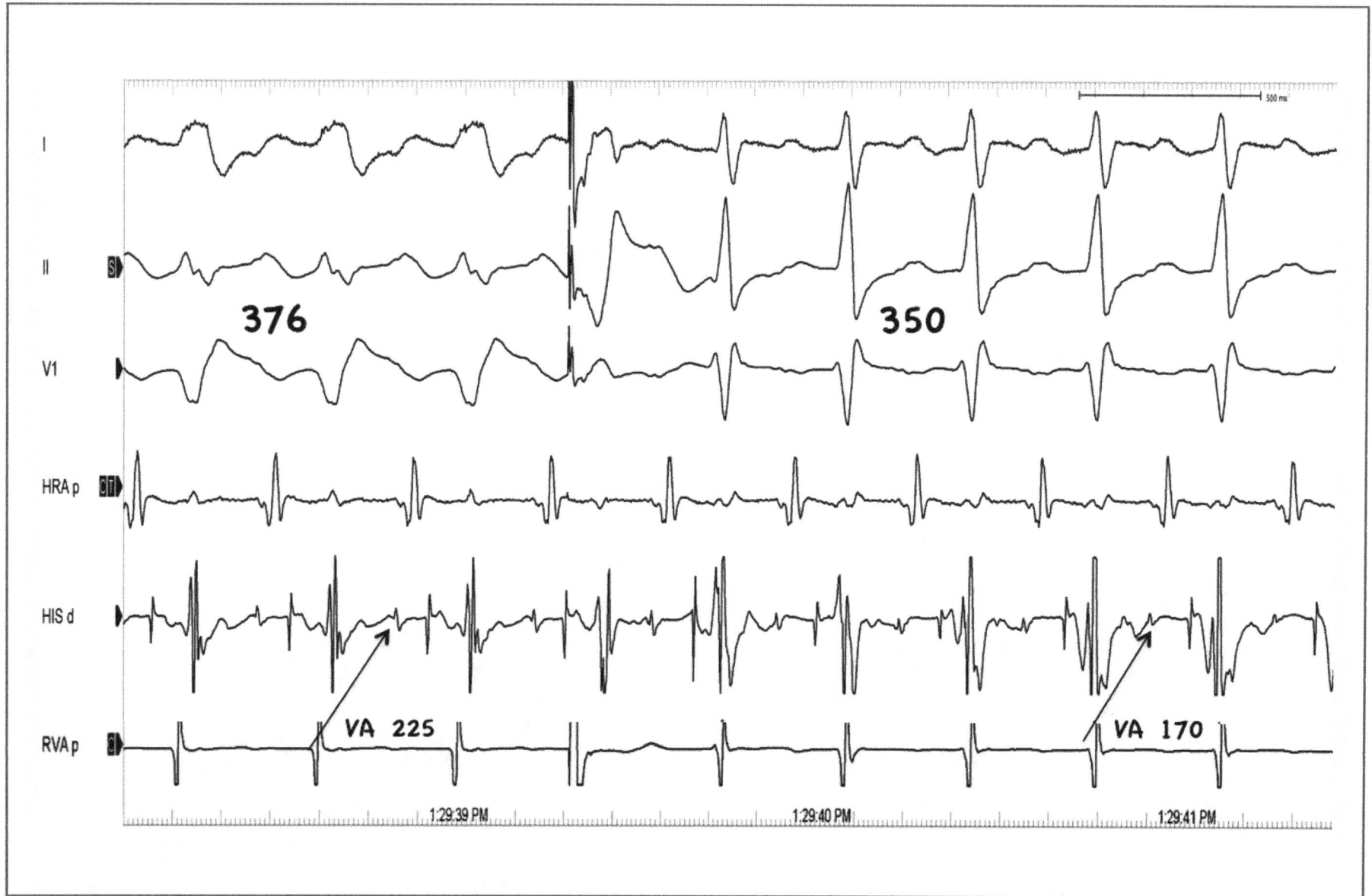

Figure 3.13

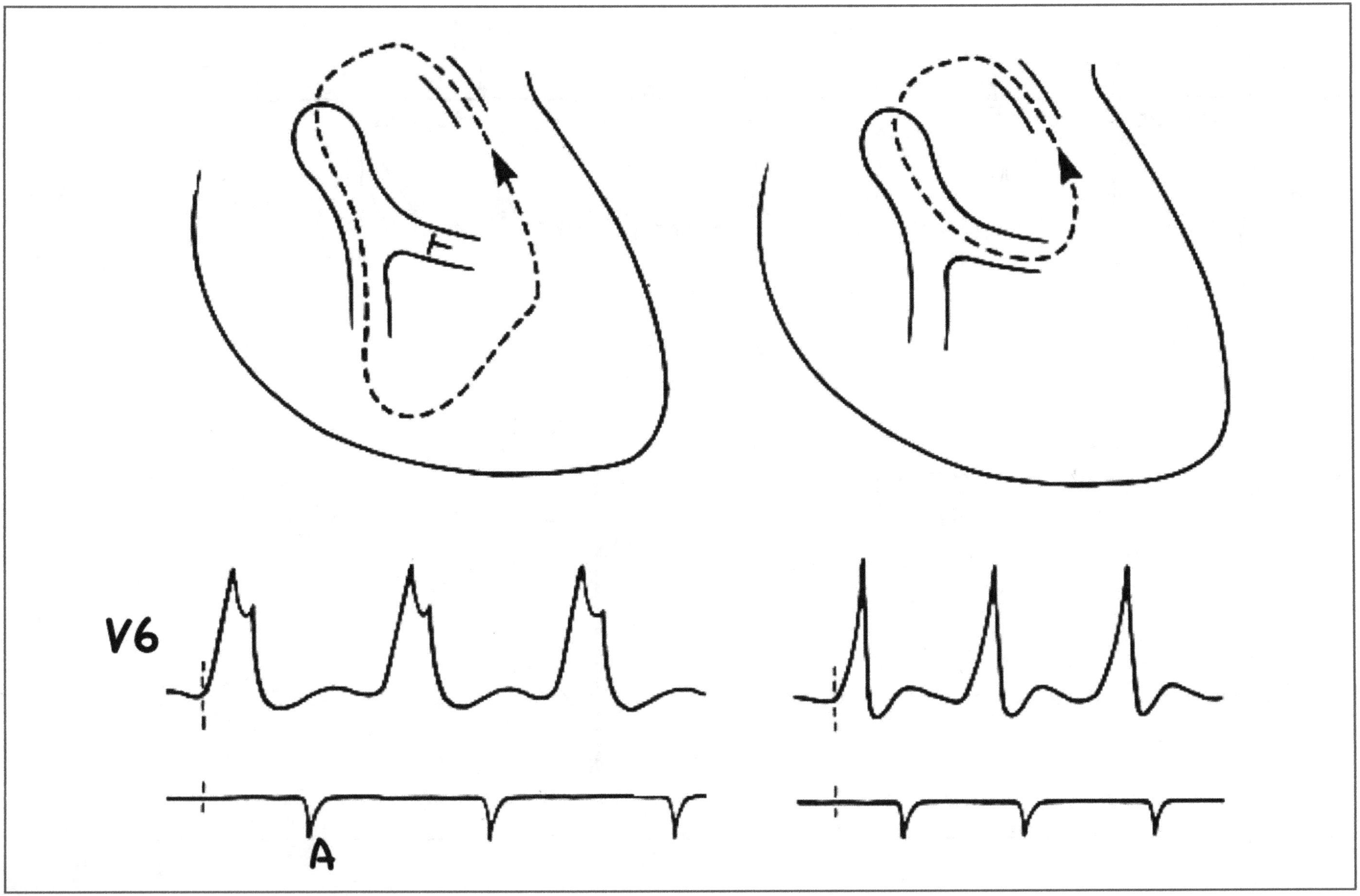

Figure 3.14

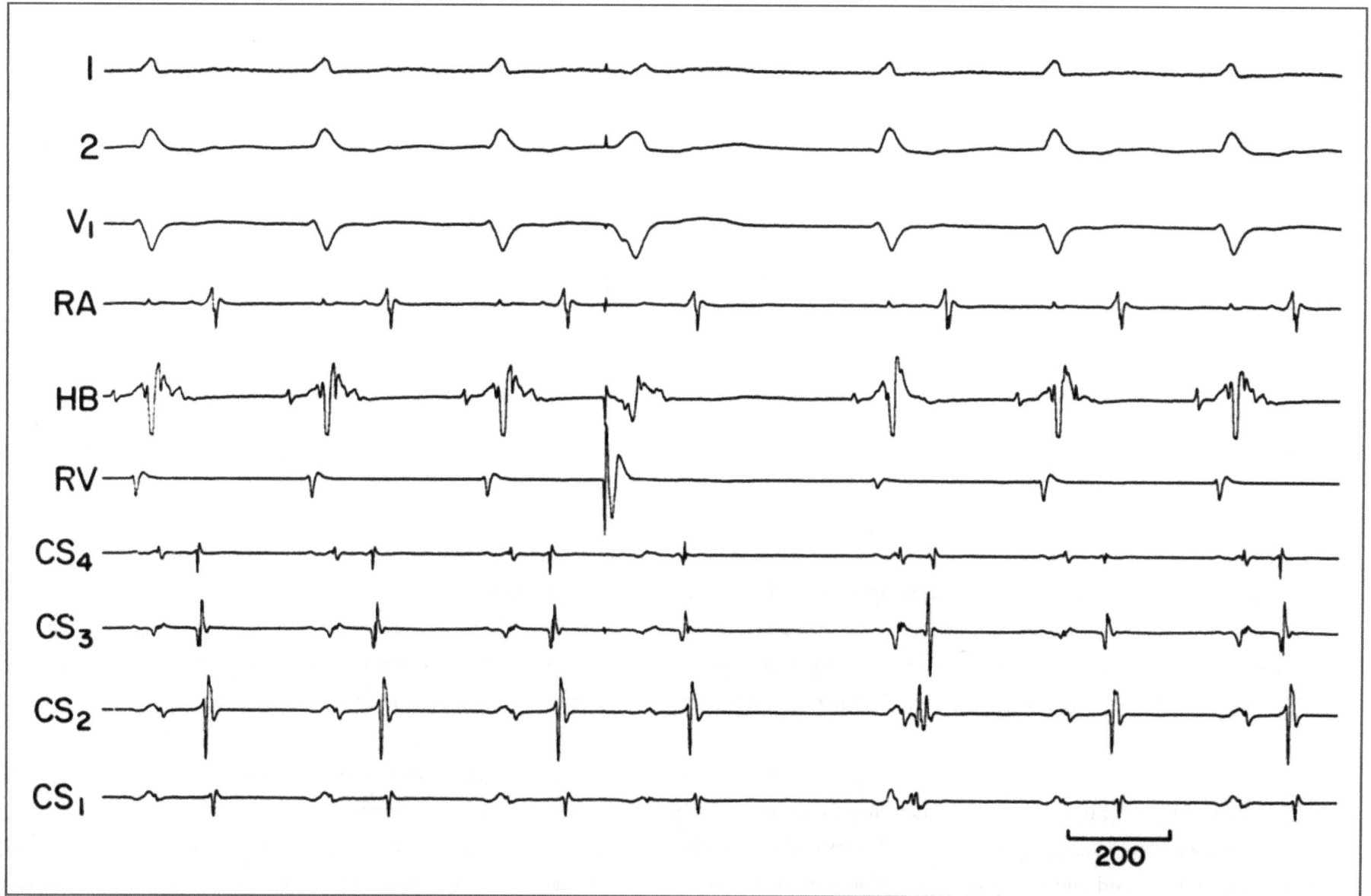

**Figure 3.15** represents an uncommon occurrence. The tachycardia at the onset has slight variability and alternates RBBB with normal QRS morphology. It has virtually simultaneous atrial and ventricular activation and *was clearly AV node reentry* as demonstrated elsewhere in the study. The His is unfortunately small but nonetheless *the PVC inserted into the cycle preexcites the next A when it should be refractory!* This is accompanied by a subtle VA change and slowing of the tachycardia. It is also noted that the PPI is now 464, suggesting that the RV pacing site is "in" the circuit, at least of the tachycardia following the PVC. This was proven to be a posteroseptal pathway elsewhere in the study. The example is a rare proof of the theoretical notion that preexciting the A when the His is refractory proves presence of an AP but *not* its participation in the initial tachycardia, in this case AVNRT.

The tachycardia at the onset in **Figure 3.16** was proven to be AVNRT elsewhere in the study and shows 2 to 1 AV block below the level of the recorded His. The spontaneous PVC may have preexcited the His ("peeling back" refractoriness), allowing it a little more time to recover excitability to the next supraventricular impulse.

Figure 3.15

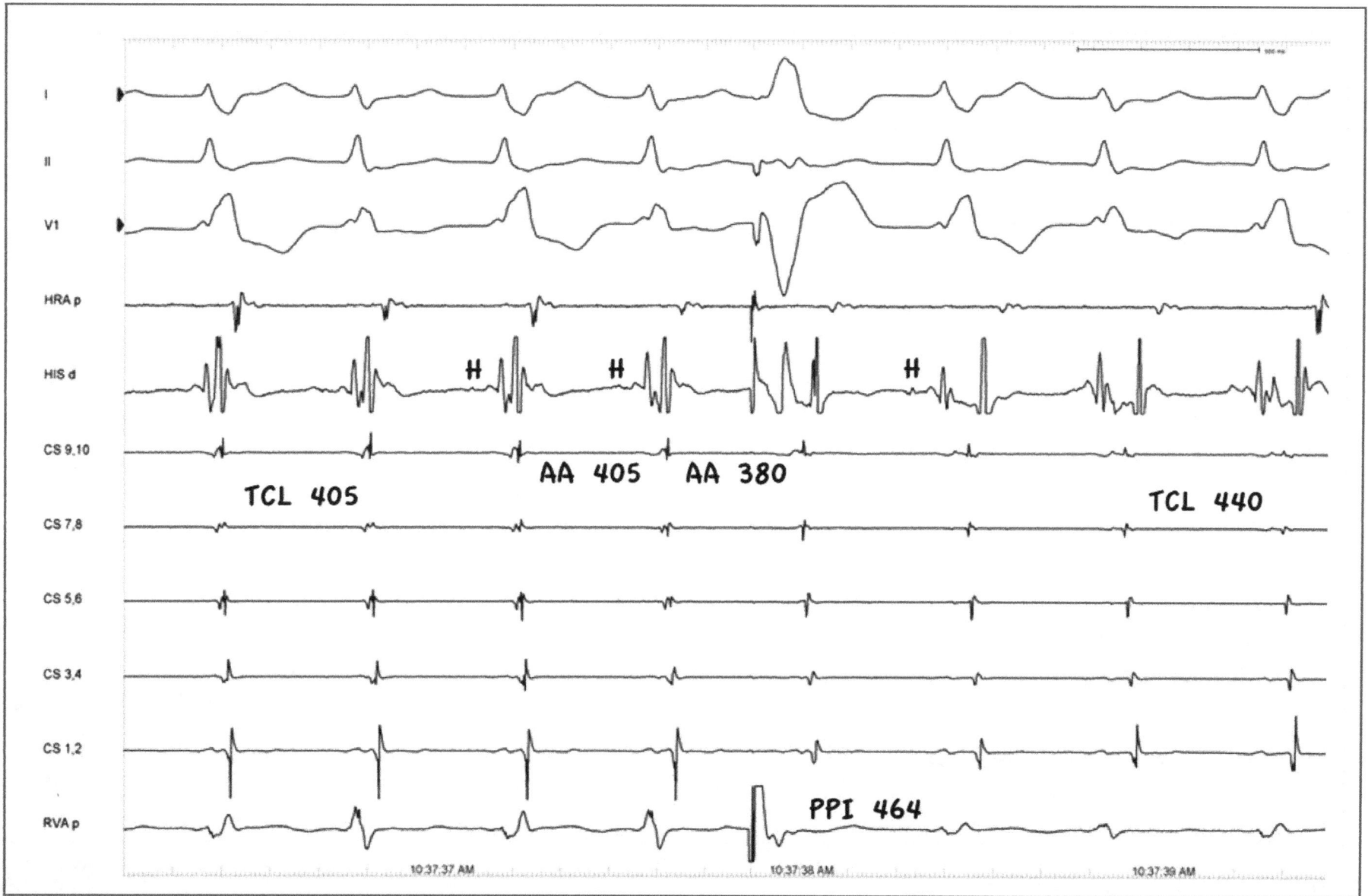

**Figure 3.16**

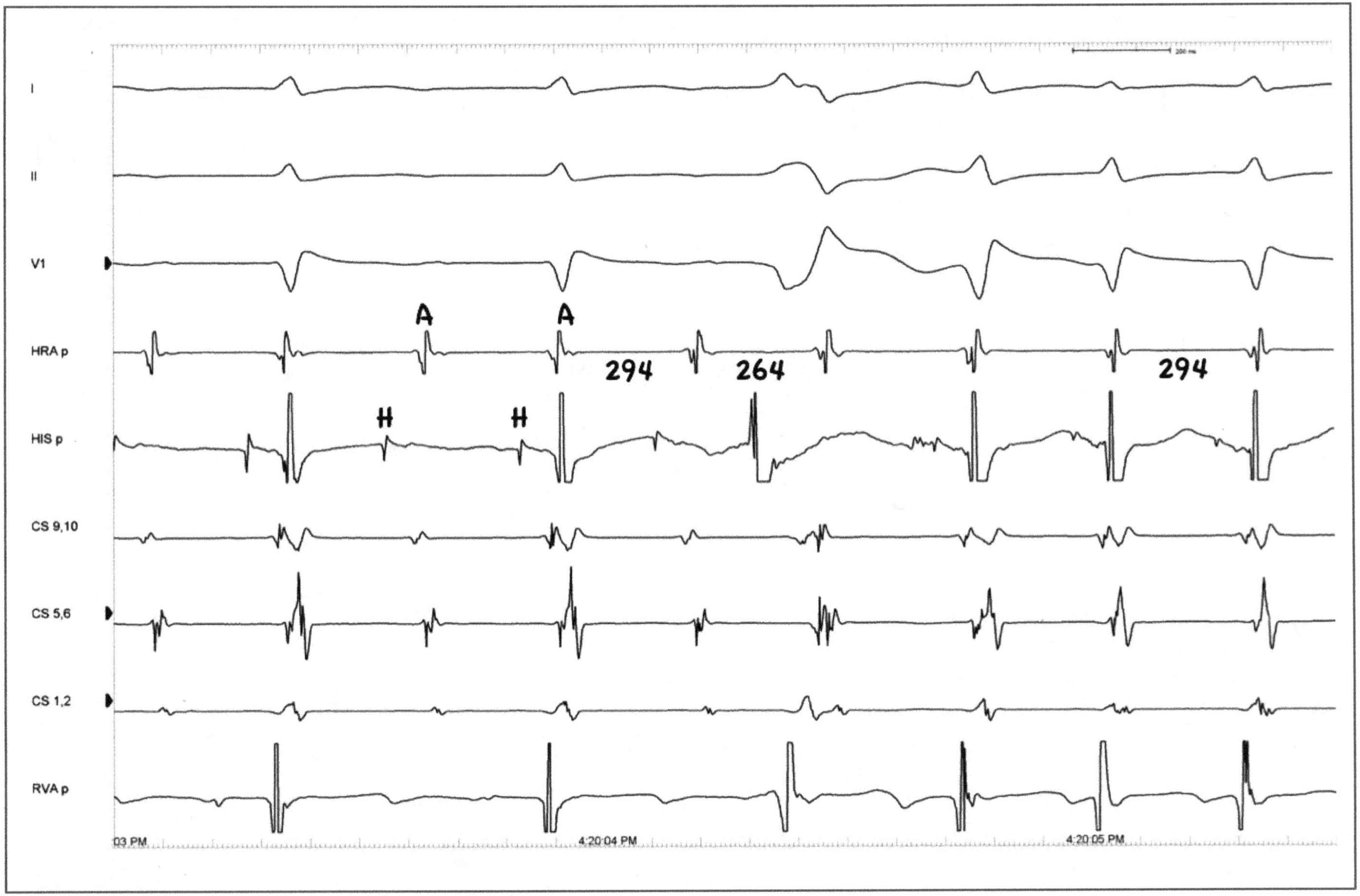

**Figure 3.17** shows a wide QRS tachycardia with a His deflection at the onset of QRS. This can only be a preexcited tachycardia or VT, and the same QRS morphology was observed during rapid atrial pacing essentially verifying preexcitation. The RV apical EGM is at the onset of QRS, most compatible with an atriofascicular AP. This was, in fact, antidromic tachycardia utilizing an atriofascicular AP in the anterograde direction and the normal AV conduction system retrogradely as illustrated schematically in **Figure 3.18**. Note from this figure that the recorded His at the onset of the QRS is actually a *retrograde* His. The proof of this comes with introducing a PAC from the RA at the time when the atrial septal region (as represented by His A EGM, proximal CS atrial EGM) is refractory. The His bundle deflection may or may not be seen to establish this, merely that the timing of the septal atrial EGM is not altered by the extrastimulus (note that the A at the CS9, 10 is "on time" when the stimulus arrives). In this case, the next preexcited QRS is *delayed* (slanted arrows, VV 370 after the atrial extrastimulus), and this can only happen if there is an AP remote from the atrial septal region. One may also look at this as a manifestation of "fusion and reset" (i.e., atrial fusion, reset of the next QRS), which one might expect with a macroreentrant tachycardia, i.e., AV reentry.

**Figure 3.19** is from the same patient. This time, the extrastimulus is a little earlier but still at a time when the atrial septum would be refractory. The tachycardia terminates without conducting to the ventricle due to block in the AP. The implications are the same as for the delay in the subsequent QRS, as observed in Figure 3.17.

Figure 3.17

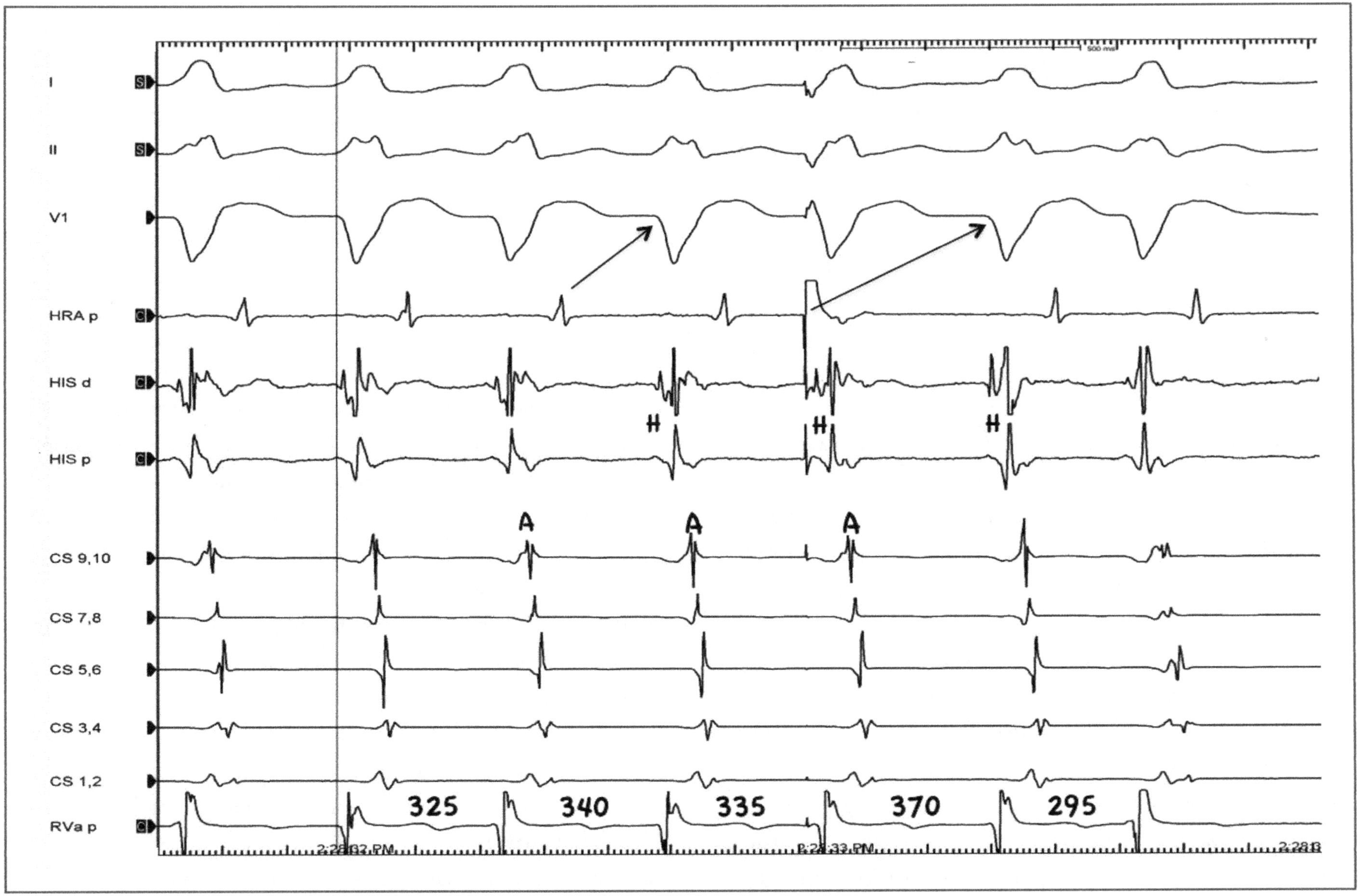

**Figure 3.18**

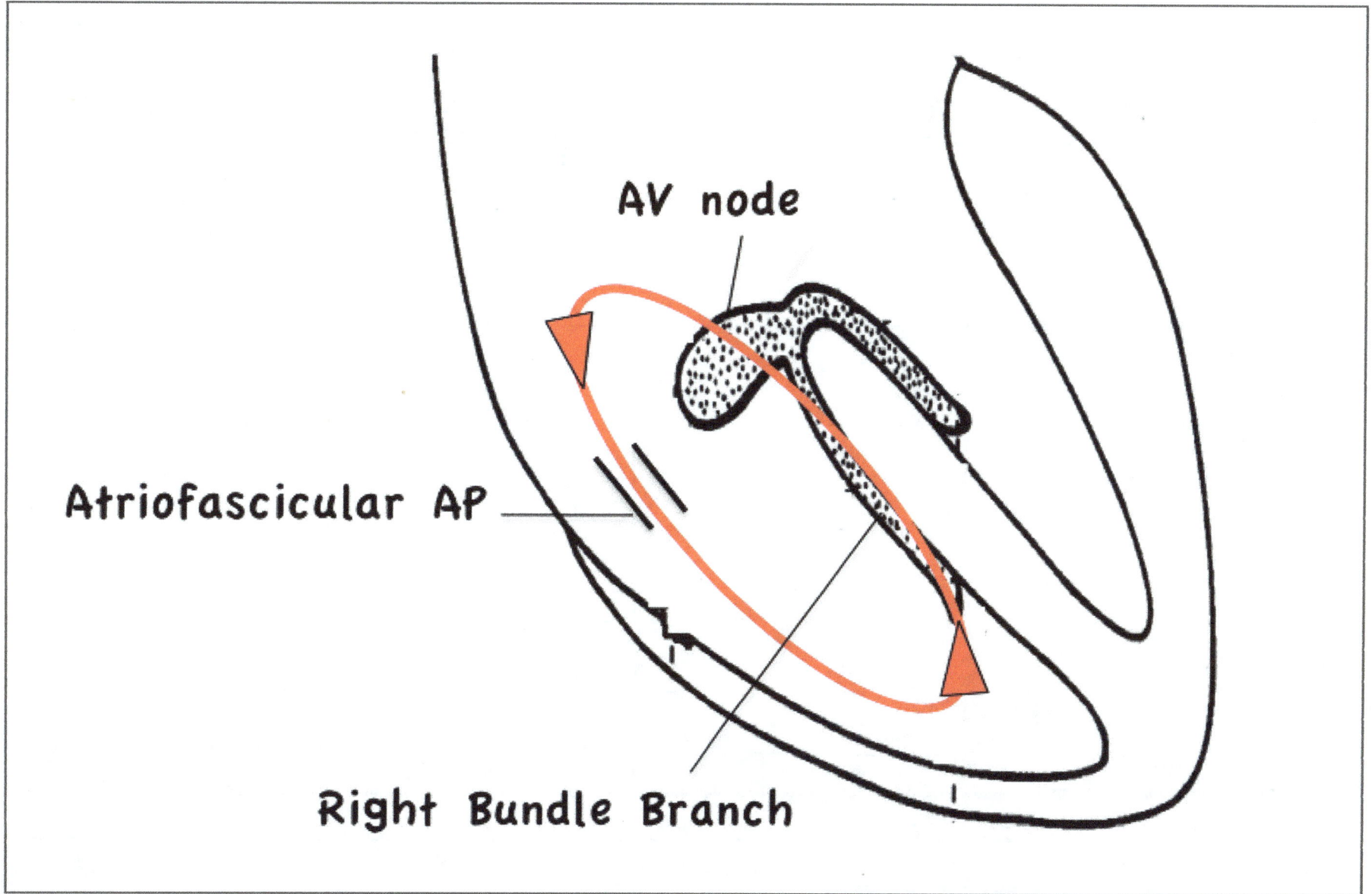

AV node

Atriofascicular AP

Right Bundle Branch

Figure 3.19

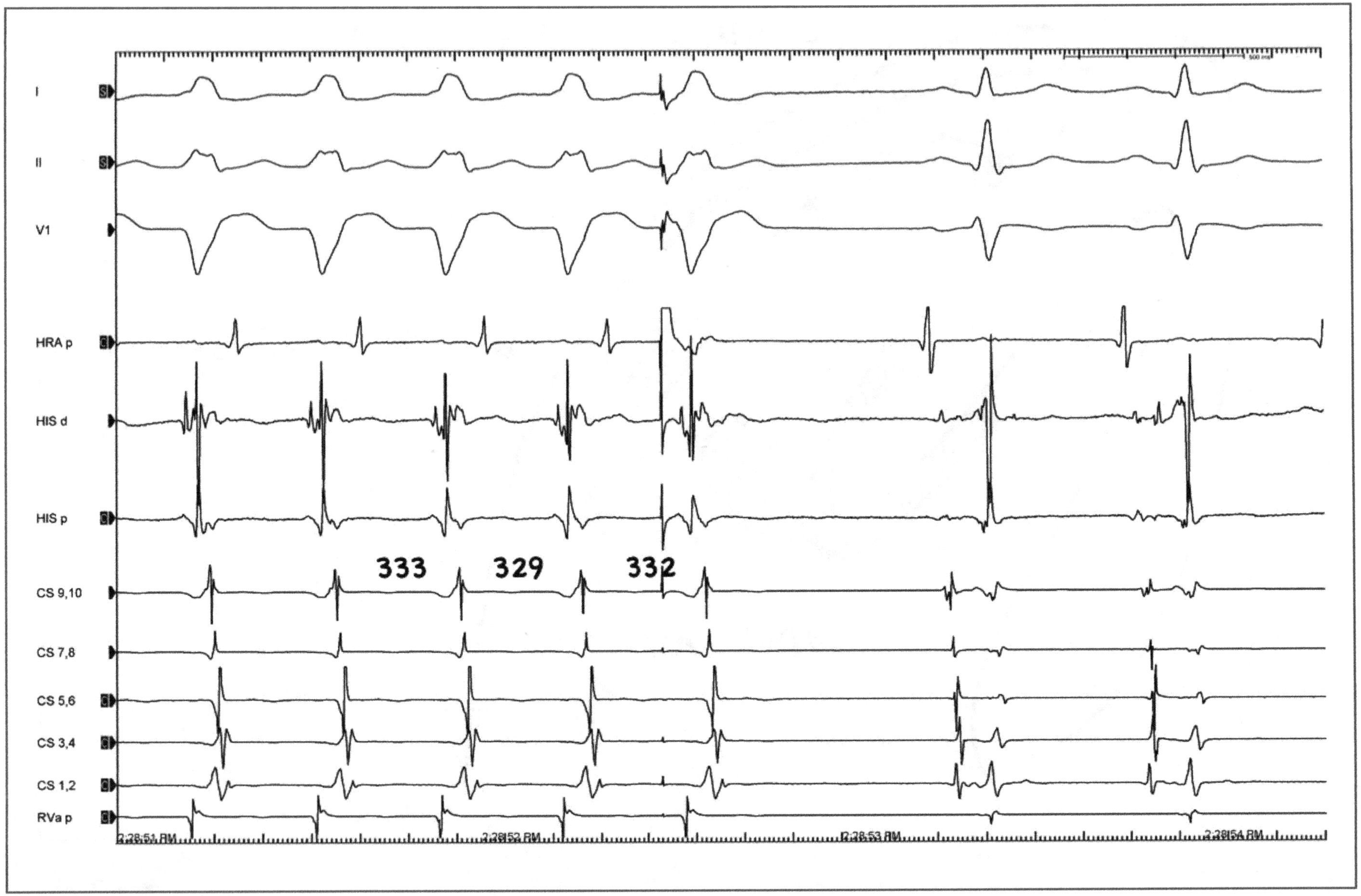

**Figures 3.20** and **3.21** illustrate a relatively common issue after slow pathway ablation in AVNRT. With atrial extrastimulus testing after the procedure, the slow pathway appears to block after a critical interval (Figure 3.20), but it is not entirely clear whether this is merely junctional tachycardia after the procedure or persistent AVNRT. Very simply, adding an S3 after the S2 generally clarifies this. If the rhythm is junctional, the S3 should advance the next QRS over a wide range of S2-S3 (as illustrated in Figure 3.21 vs. 3.20), whereas it would not be possible to advance the next QRS if it were AVNRT over a slow pathway except perhaps for a very brief interval as illustrated in the subsequent figures. This is because the AV node would be refractory for most of the PR interval when conduction is going over a slow pathway, whereas this should not be the case in the diastolic interval between QRS complexes in junctional rhythm.

Figure 3.20

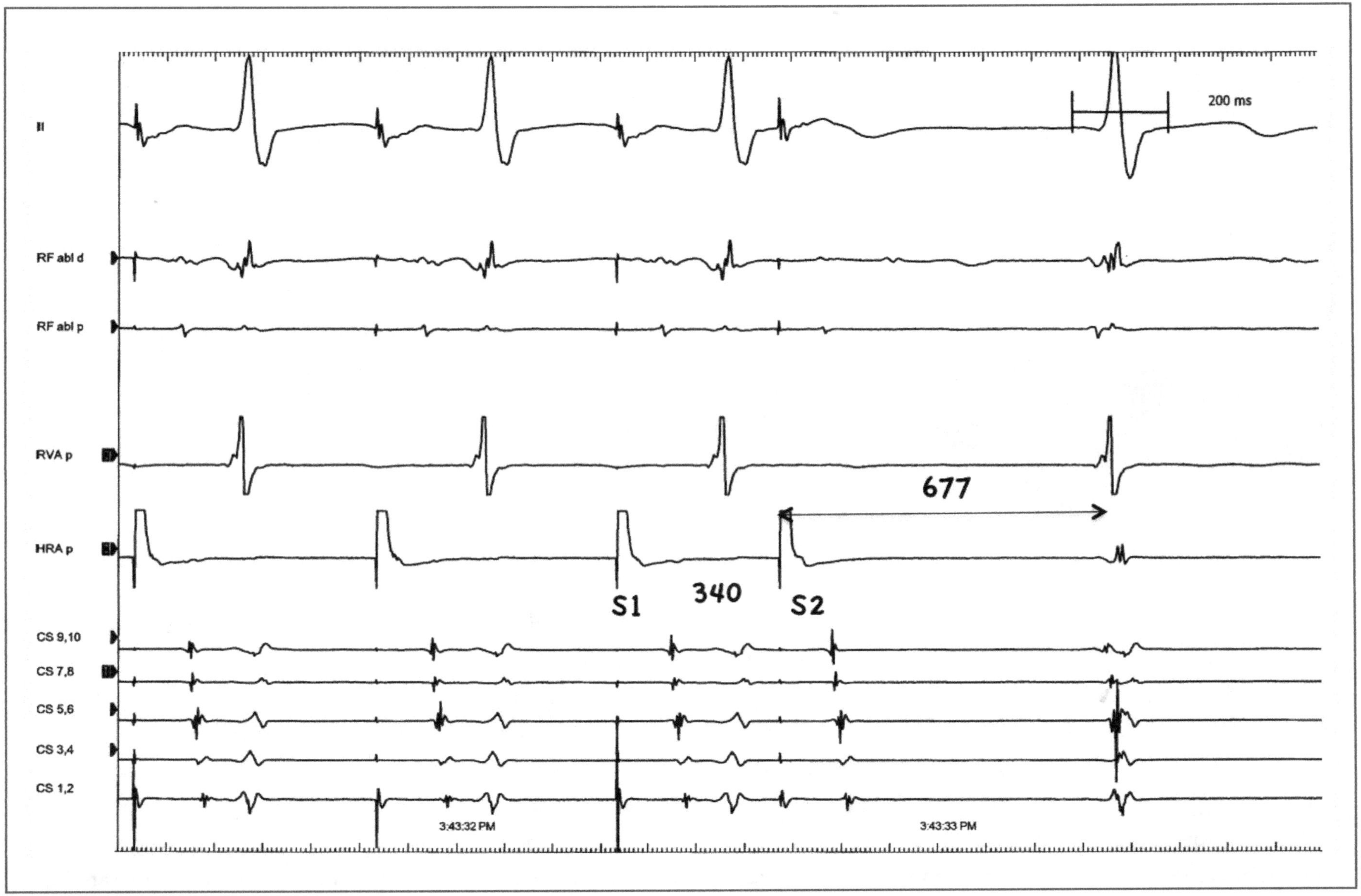
II
200 ms
RF abl d
RF abl p
RVA p
677
HRA p
S1
340
S2
CS 9,10
CS 7,8
CS 5,6
CS 3,4
CS 1,2
3:43:32 PM
3:43:33 PM

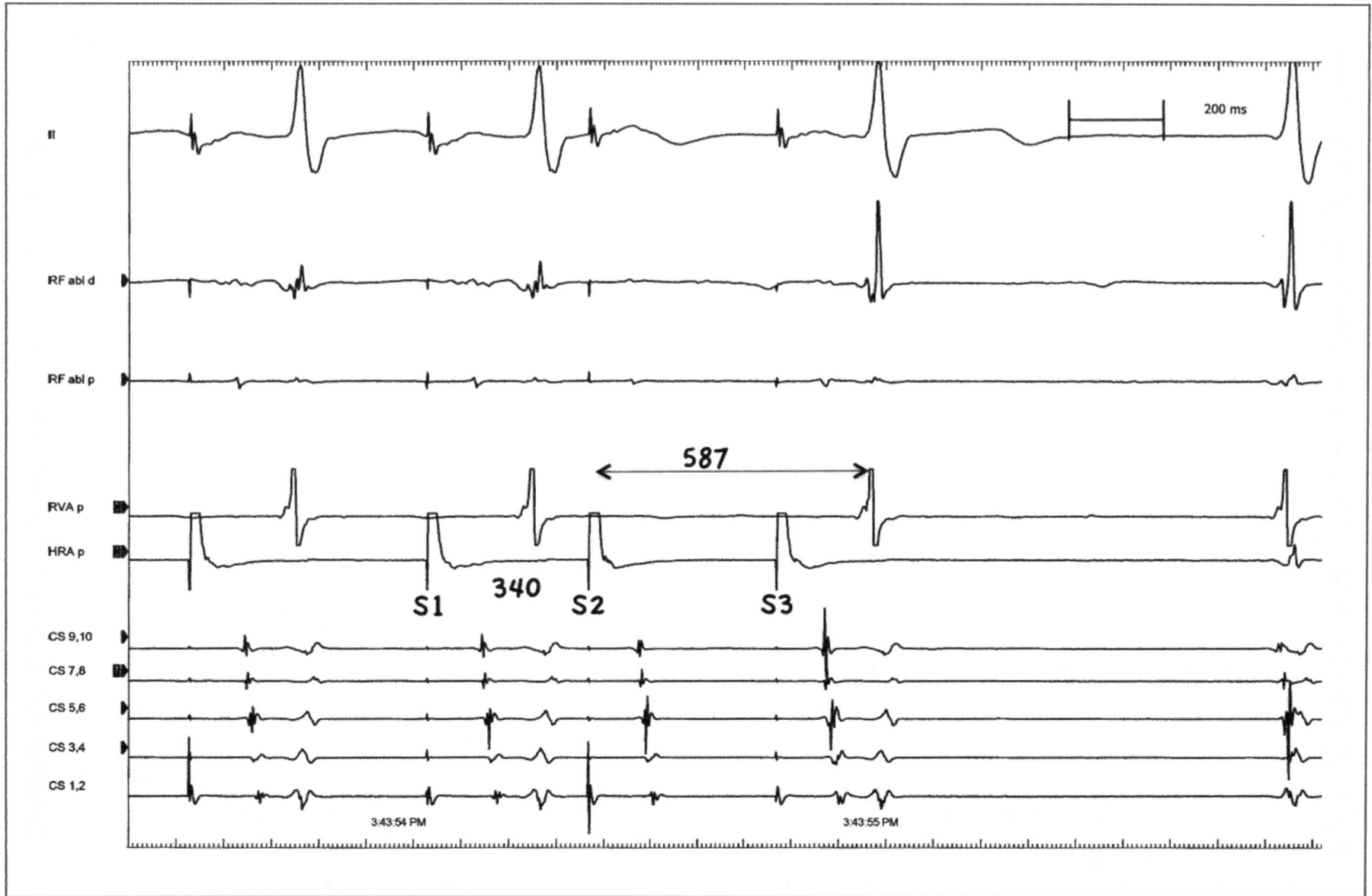

Figure 3.21

Continuing the same theme in **Figure 3.22**, the issue is again the differential diagnosis between junctional tachycardia (JT) and AVNRT. An atrial extrastimulus is introduced when the His should be refractory and this should therefore have *no* influence on JT. However, as we see, the *subsequent* cycle was advanced (390 to 360) and this could only occur if the atrial extrastimulus entered the excitable gap in AVNRT after the slow pathway recovered excitability and advanced the *next* cycle. **Figure 3.23** was recorded from the same patient, only with the extrastimulus slightly earlier. This could not be JT, of course, which would not be expected to terminate with a single atrial extrastimulus. It is, however, reasonably explained by postulating that the extrastimulus engaged the *fast pathway* anterogradely and caused block in this when the slow pathway came around the next time. Alternately, the extrastimulus was early enough to encounter refractoriness in the slow pathway and terminate the tachycardia.

Figure 3.22

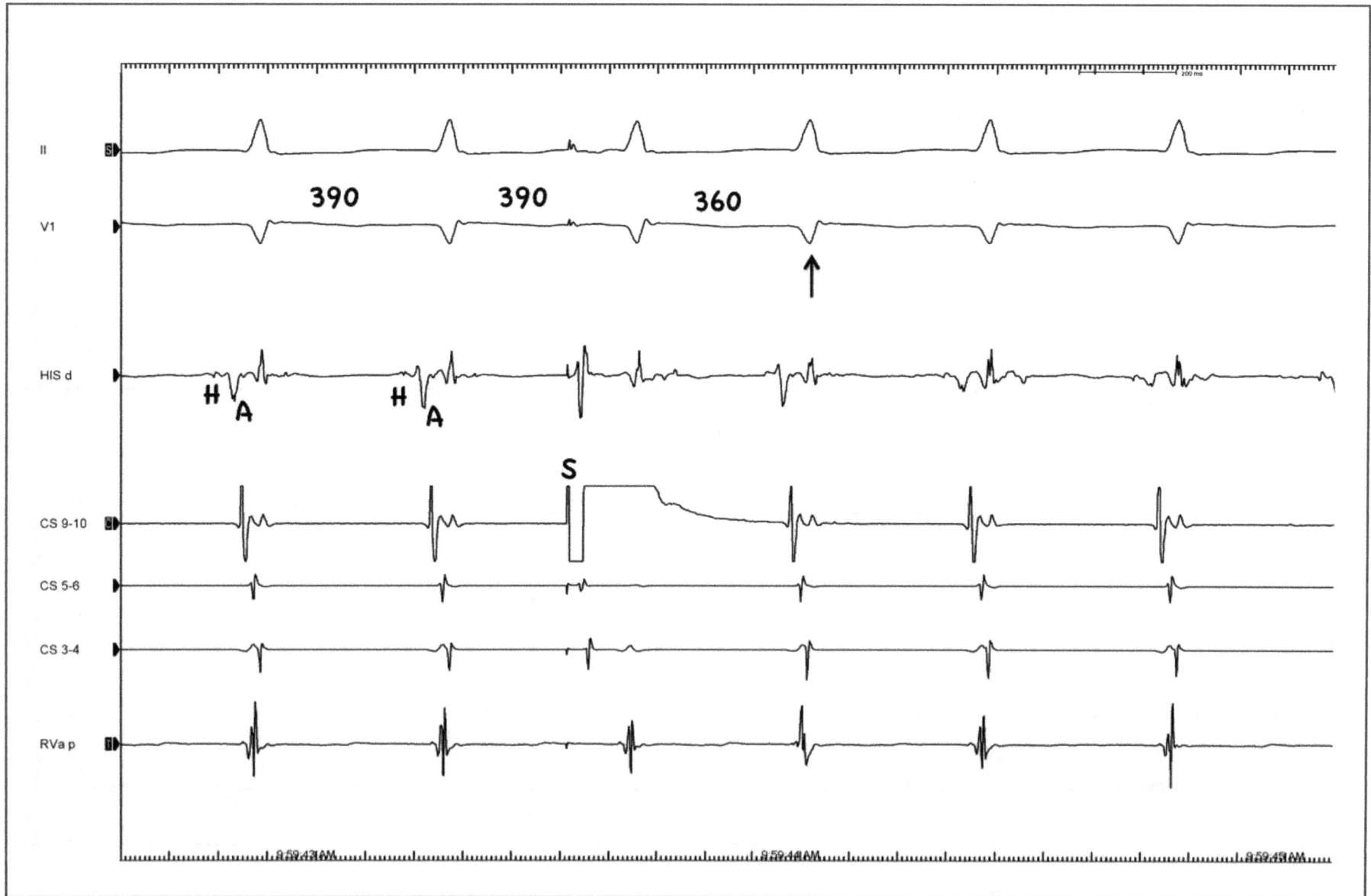

**Figure 3.23**

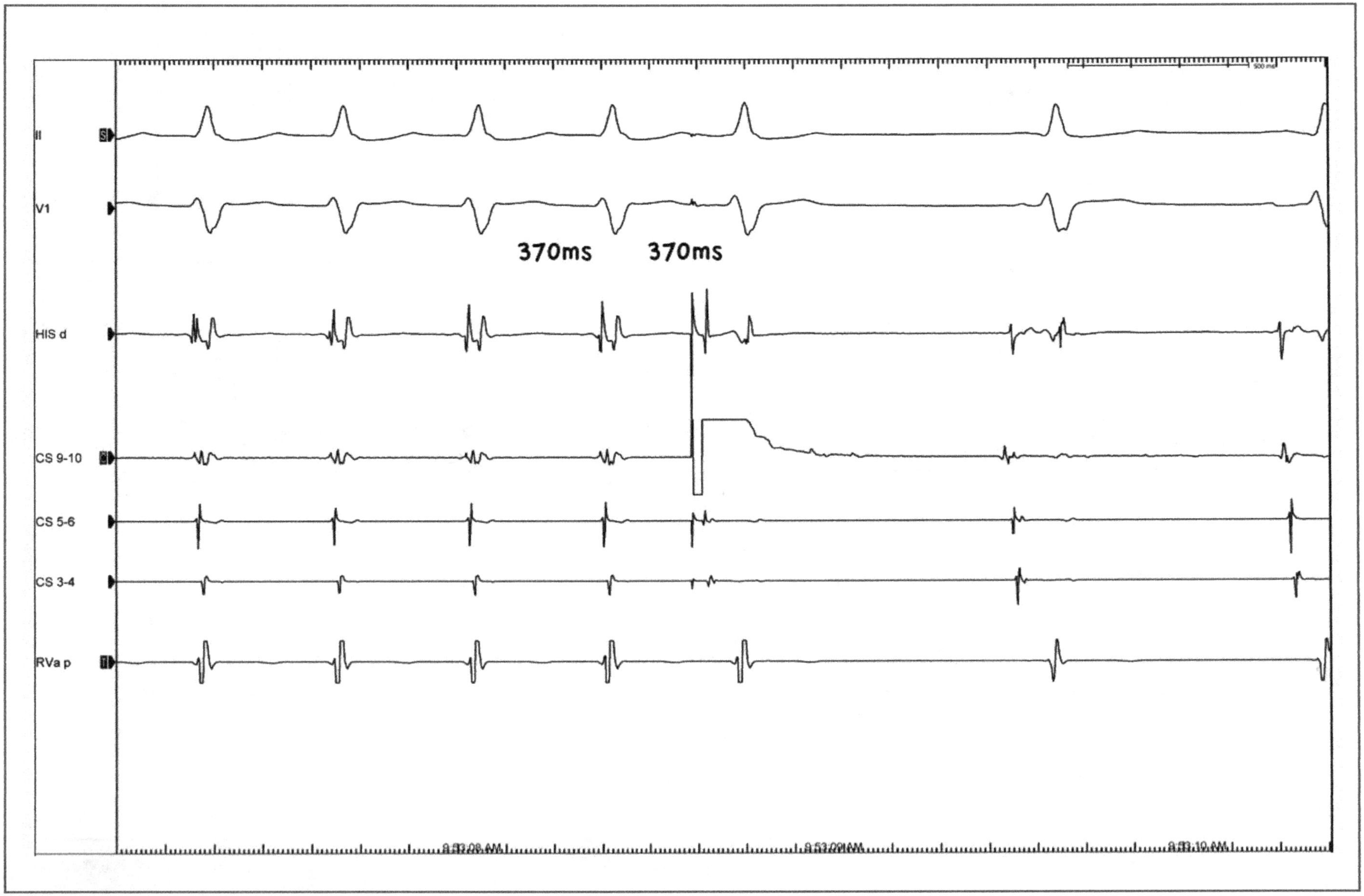

The theme is continued further in **Figures 3.24** and **3.25** with a little more complexity. In the former, the atrial extrastimulus occurs during His refractoriness and would not influence JT. Yet it advances the subsequent cycle as per the previous examples, and this is only compatible with AVNRT. Figure 3.25 in the same patient shows a slightly earlier extrastimulus and this advances *both* the next *and* the subsequent cycle! In this example, the His appears to be advanced slightly. This is most compatible with a "2 for 1" phenomenon where a single extrastimulus captures both the fast and the slow pathway and tachycardia continues thereafter. To support this, "2 for 1" AV nodal response was observed elsewhere in the study.

## Figure 3.24

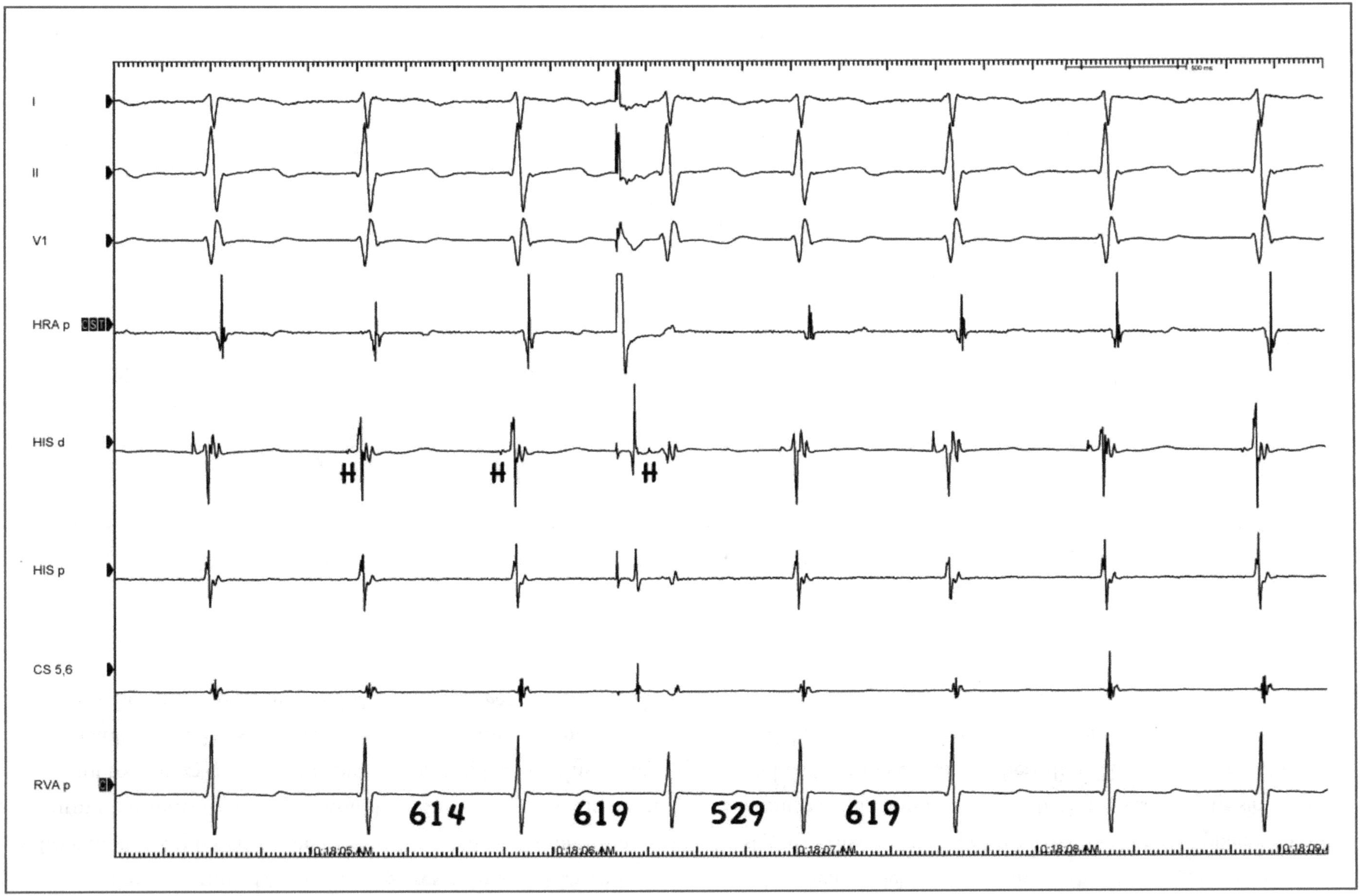

**Figure 3.25**

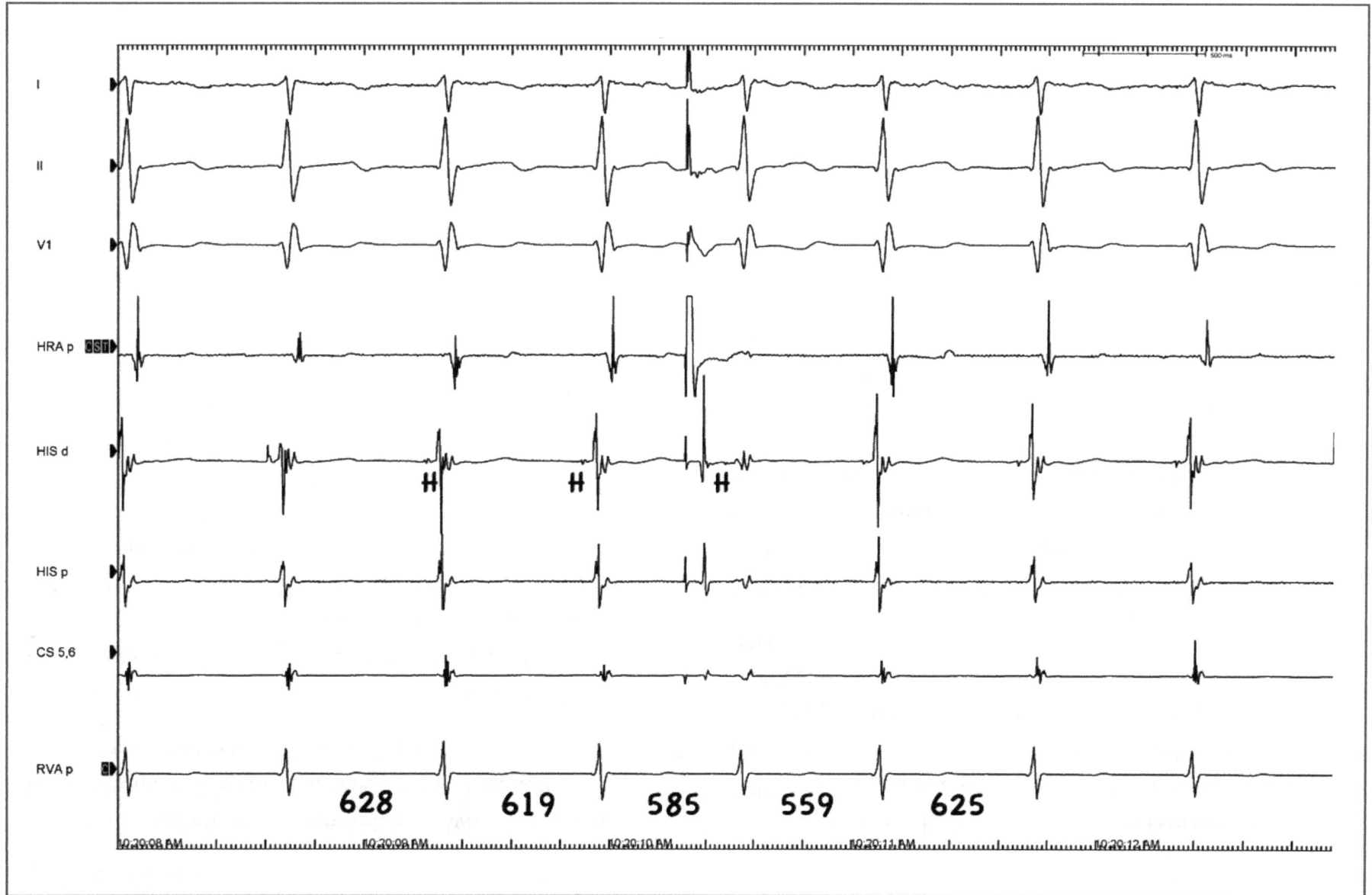

## Suggested Readings

1.  Gallagher JJ, Gilbert M, Svenson RH, et al. Wolff-Parkinson-White syndrome. The problem, evaluation, and surgical correction. *Circulation*. 1975;51:767–785.

2.  Michaud GF, Tada H, Chough S, et al. Differentiation of atypical atrioventricular node re-entrant tachycardia from orthodromic reciprocating tachycardia using a septal accessory pathway by the response to ventricular pacing. *J Am Coll Cardiol*. 2001;38:1163–1167.

3.  González-Torrecilla E, Almendral J, Garcia-Fernandez J, et al. Differences in ventriculoatrial intervals during entrainment and tachycardia: a simpler method for distinguishing paroxysmal supraventricular tachycardia with long ventriculoatrial intervals. *J Cardiovasc Electrophysiol*. 2011;22:915–921.

4.  Padanilam BJ, Manfredi JA, Steinberg LA, et al. Differentiating junctional tachycardia and atrioventricular node re-entry tachycardia based on response to atrial extrastimulus pacing. *J Am Coll Cardiol*. 2008;52:1711–1717.

# Overdrive Pacing and Entrainment:

## Supraventricular Tachycardia

A lthough atrial overdrive pacing for supraventricular tachycardia (SVT) can be useful, ventricular overdrive pacing is much more commonly used and generally more valuable. Pacing is begun at a cycle length 20 ms or so shorter than the tachycardia cycle length (TCL), the pacing rate being fast enough to appreciate a difference but hopefully not fast enough to cause conduction delay in the circuit that could confound the measurements. The pacing should last for long enough to achieve a steady state, usually 10 to 20 seconds.

Three responses may usually be anticipated:

1. No VA conduction with the atrial rate continuing unabated, generally confirming a diagnosis of atrial tachycardia (AT).

2. Overdrive suppression of the tachycardia with retrograde VA conduction and return of AT after pacing is discontinued. This is the "VAAV" response that will be illustrated.

3. Entrainment of the tachycardia or direct acceleration of the tachycardia while the pacing is continued, with resumption of the tachycardia rate when the pacing stops. This is the "VAV" response that will be illustrated.

## Avoidance of Common Mistakes: The Entrainment Checklist

There are many pitfalls in overdrive pacing and a checklist approach can prove useful to avoid many of these. **Figure 4.1** is a straightforward example of an AT but serves to illustrate our checklist. The tachycardia in the right half of Figure 4.1 is SVT with eccentric atrial activation sequence (RA first).

1. Measure cycle length (CL) of the tachycardia and note its regularity. Obviously, the more CL irregularity, the more difficult it is to draw conclusions. In this example, the tachycardia is regular with CL 390 ms.

2. Look at the left part of the tracing and verify that pacing has accelerated to the pacing rate consistently, at least for the last few cycles of the pacing. In our example, the pacing CL is 370 ms and the tachycardia has been similarly accelerated.

3. Identify the last electrogram (EGM) that has been accelerated to the pacing rate. *This is a critical step in interpretation!* In our example, the dense arrow indicates the last accelerated atrial EGM. The SVT then resumes with an initial atrial depolarization beginning at the RA EGM that is not directly linked to the last paced ventricular EGM. It is thus obvious that this is a VAAV response and must be AT. Note also that the atrial activation sequence is different during pacing and tachycardia, reflecting the fact that the AT has not been "entrained" but actually overdriven by a faster atrial rate resulting from retrograde conduction during ventricular pacing with resumption of atrial tachycardia after pacing is stopped.

4. Finally, measure the postpacing interval (PPI) at the ventricular pacing site (702 ms in our example), the SA interval during pacing, and the VA interval during tachycardia. These are *not* relevant however in this example, since it is an *AT*

**Figure 4.1**

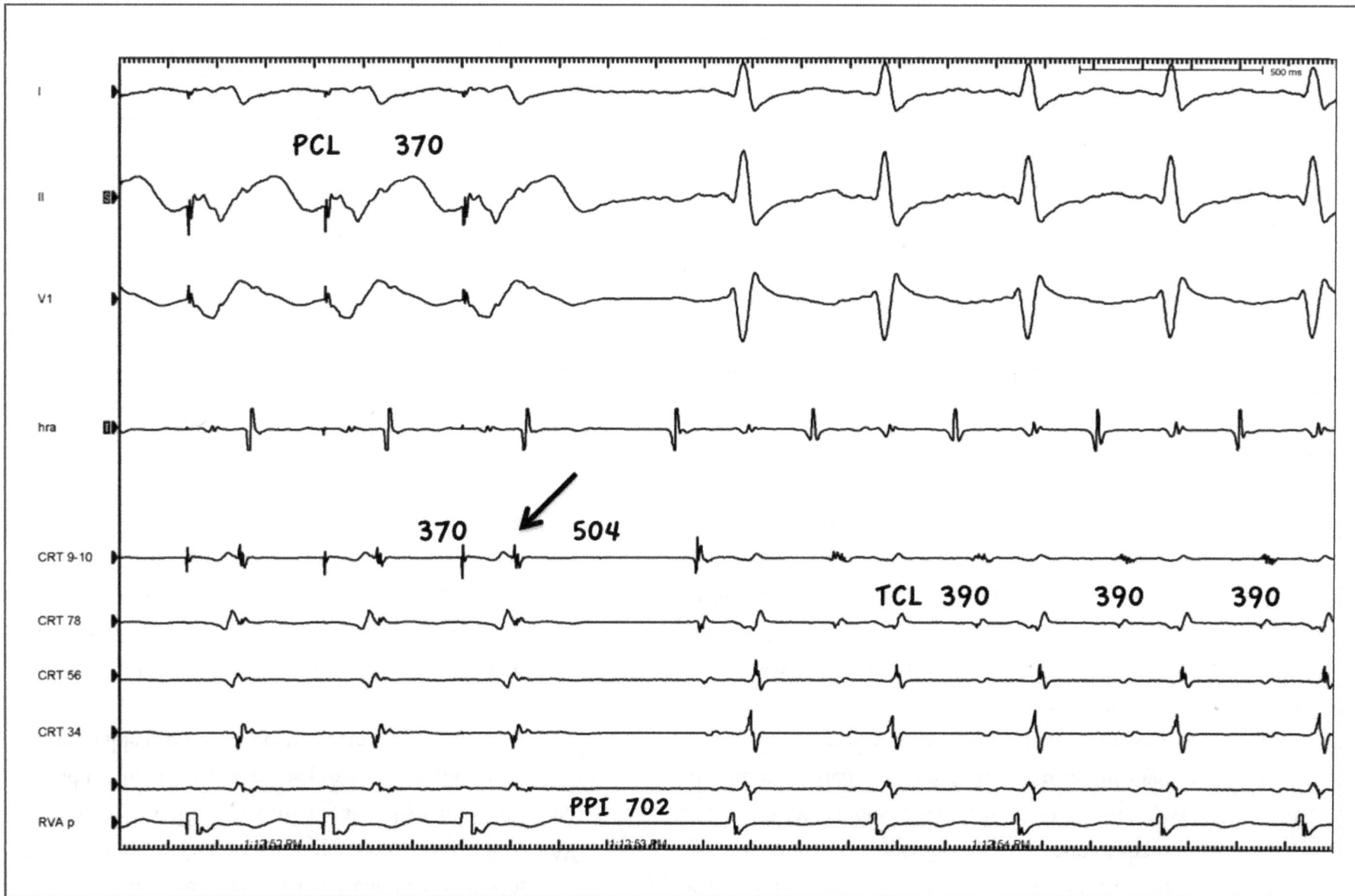

(VAAV) and there is no actual VA conduction during tachycardia. The intervals become more relevant for a VAV response where the tachycardia is actually entrained from the V and the intervals reflect the proximity of the pacing site to the excitable gap of the reentrant circuit.

The checklist quickly facilitates the correct interpretation of **Figure 4.2**. This is an SVT with a one-to-one AV relationship and eccentric atrial activation sequence, and can be only orthodromic atrioventricular (AV) reentry over a left lateral accessory pathway (AP), a left AT, or, theoretically, a junctional tachycardia with a left lateral AP present but not necessary to the tachycardia mechanism (i.e., a bystander). A superficial examination of the tracing will suggest a VAV response after termination of pacing with the last A influenced by pacing followed directly by a QRS with resumption of tachycardia. However, making the measurements more carefully as per the entrainment checklist quickly shows that the TCL is slightly irregular with the result that one can't say with confidence that the tachycardia has accelerated to the pacing rate (note the difference in VA interval between the first ventricular paced beat seen and the last). Thus, this is nondiagnostic and must be repeated using a slightly faster drive cycle length for ventricular pacing.

## Practical Usage of Overdrive Pacing/Entrainment

**Figure 4.3** shows a regular SVT with virtually simultaneous A and V activation with the realistic differential diagnosis being atrioventricular nodal reentrant tachycardia (AVNRT), AT conducting over a slow AVN pathway, or junctional tachycardia. Ventricular pacing has captured the right ventricle resulting in a fully paced QRS morphology in the first 4 beats of the tracing. The atrial rate has similarly accelerated to the pacing rate due to VA conduction, essential to interpretation of the PPI. The last atrial EGM advanced to the pacing rate (arrow) immediately follows the last paced QRS complex. Also note that the atrial activation sequence during pacing is identical to that of tachycardia as would be expected with entrainment of AVNRT or atrioventricular reentrant tachycardia (AVRT). This last accelerated A is followed by the QRS (His bundle not recorded but H deflection would be expected in front of V) indicating that the advanced A directly causes the next V (VAV). This indicates that the mechanism must be AVNRT or AVRT. This is now obviously AVN reentry as the VA interval during tachycardia is too short to be AV reentry, but this is verified by measurement of the PPI, in comparison to the TCL. In this case, the PPI is 440 and the TCL 310, thus a difference of 130 ms. The PPI–TCL greater than 115 ms indicates that the pacing site is "far" from the tachycardia circuit and not consistent with AVRT over the usual type of septal AP. We will discuss in subsequent examples that this does not necessarily exclude unique APs with long conduction time and decremental properties.

**Figure 4.2**

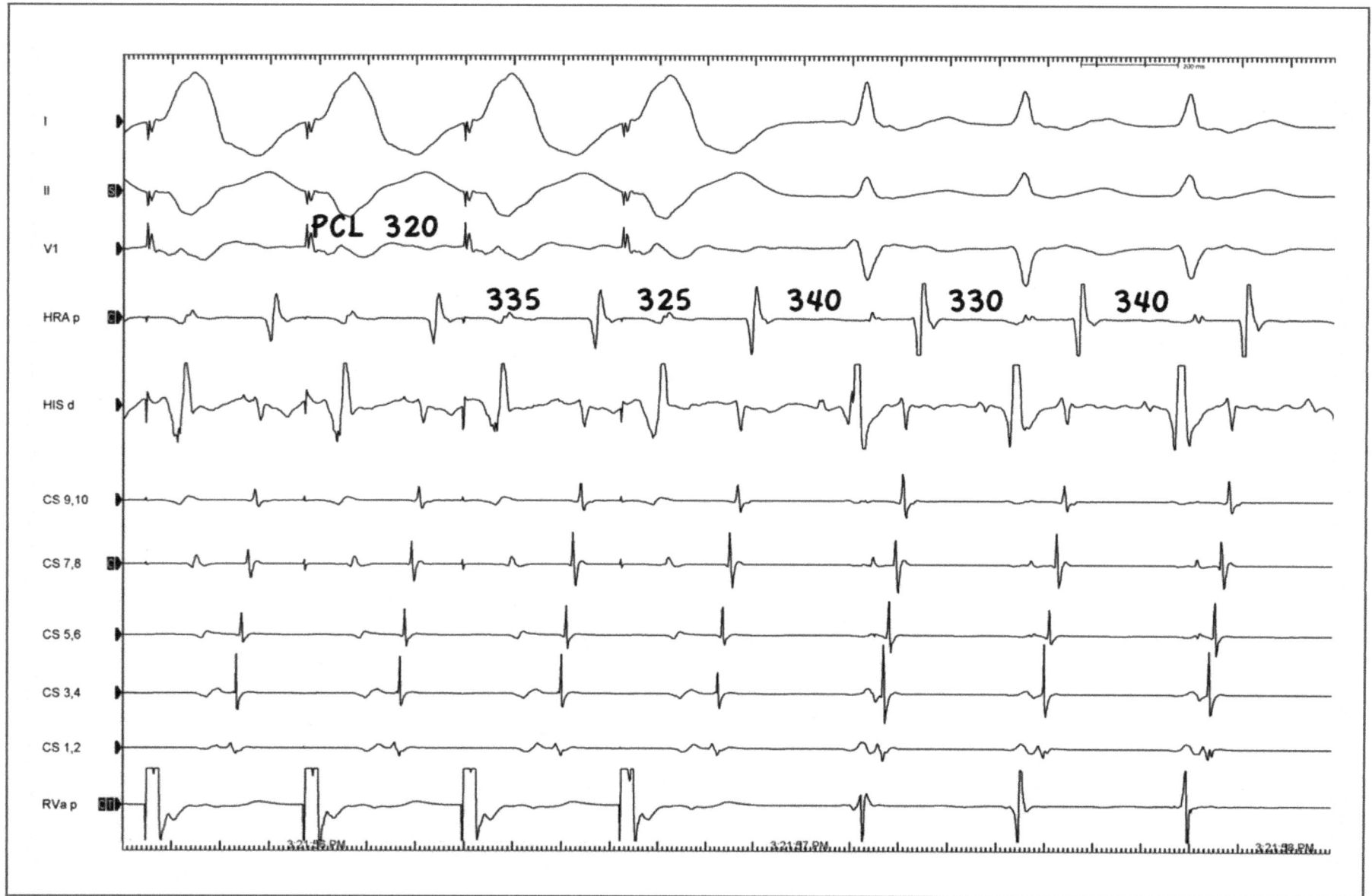

**Figure 4.3**

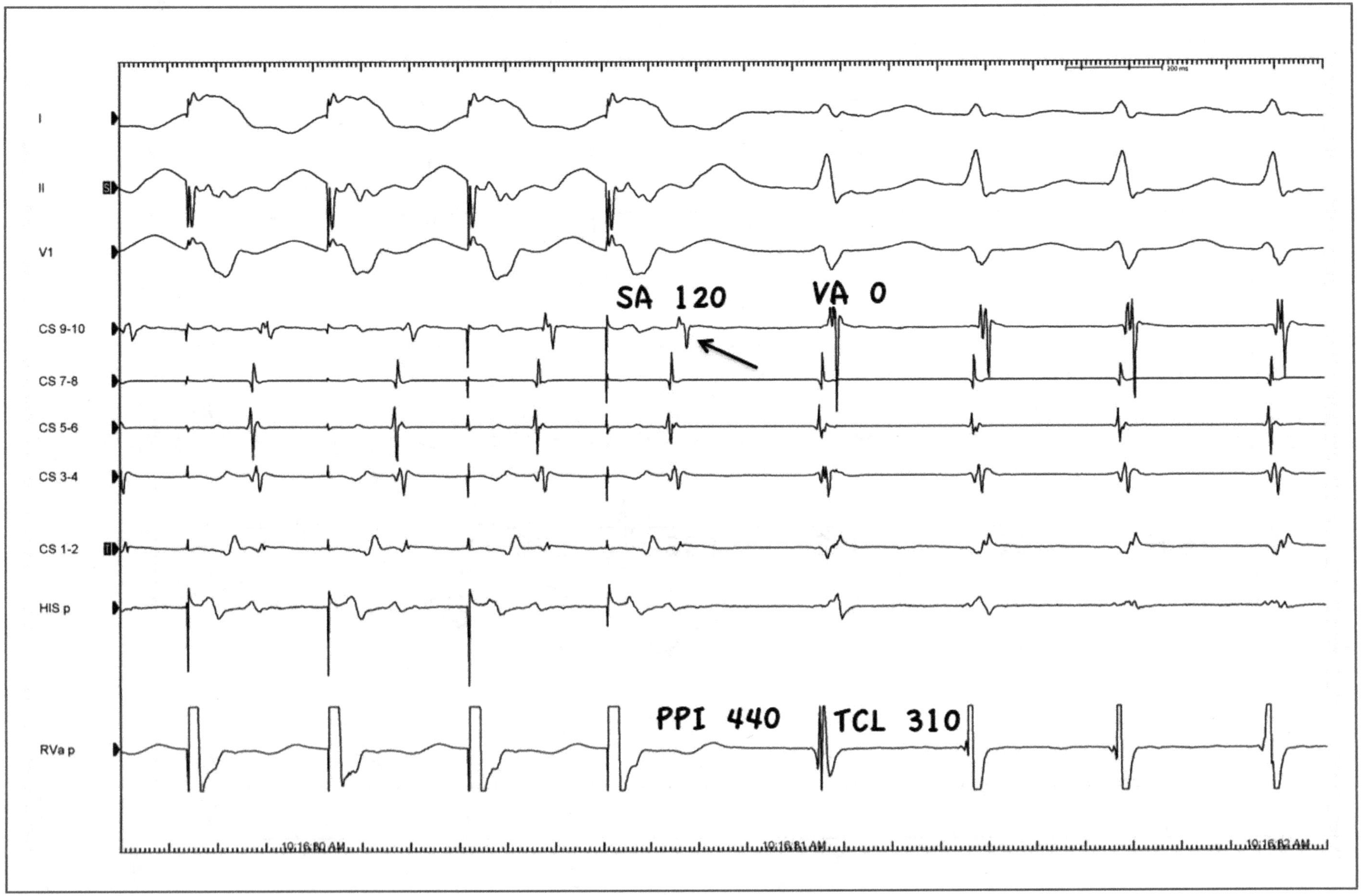

This is further supported by comparison of the SA interval (120) during pacing to the VA interval during tachycardia (0). This measurement reflects the retrograde limb of this circuit, eliminating the potential for decrement in the AV node to confound the PPI measurement. In this case, the large difference between SA and VA (120 ms) is greater than 85 ms, further verifying AVNRT as the tachycardia mechanism.

**Figure 4.4** illustrates an interesting regular SVT with a 1:1 AV relationship. The atrial activation is a little strange in that the His EGM A (His activation not seen) is earliest, yet the coronary sinus (CS) atrial activation shows the distal CS atrial EGM to be earlier than the proximal atrial EGM. We are not sure exactly where the CS catheter is located and a "septal" pathway may cause an apparent reverse CS atrial activation if the catheter is advanced too far. With this uncertainty, we can nonetheless proceed with overdrive pacing.

Pacing from the RV apex (RVA) captures and drives the first 3 QRS complexes and advances the atrial EGM to the pacing rate. The last advanced atrial EGM is identified immediately following the third paced QRS complex and is in turn followed by a QRS with resumption of the tachycardia.

The atrial activation sequence during pacing is identical to the tachycardia, consistent with entrainment of SVT rather than overdrive of an AT. The VAV pattern identified is consistent with either AVNRT or AVRT but not AT. The difference between PPI (470 ms) and TCL (420 ms) is short at 50 ms, indicating that the RVA pacing site is relatively close to the excitable gap of the reentrant circuit. A PPI–TCL less than 115 ms when pacing from the RVA is widely accepted as indicating AVRT (see Suggested Readings #3). This is further supported by comparing the SA (240 ms) to the VA time during tachycardia (205 ms), with a difference of 35 ms. A difference of 85 ms or less is indicative of AVRT.

Not only does this maneuver diagnose AV reentry but gives a reasonable clue to AP location. In this instance, the very short PPI–TCL of 50 ms suggests that the RV pacing site is very close to the circuit most compatible with a right or septal AP. One might also note here that the AH interval following the last entrained A did not prolong relative to that during tachycardia, verifying that a prolonged AH does not prolong and confuse the PPI measurement. When this happens, the PPI is artificially long and can be "corrected" by subtracting the increment in the AH interval in the postpacing cycle.

**Figure 4.4**

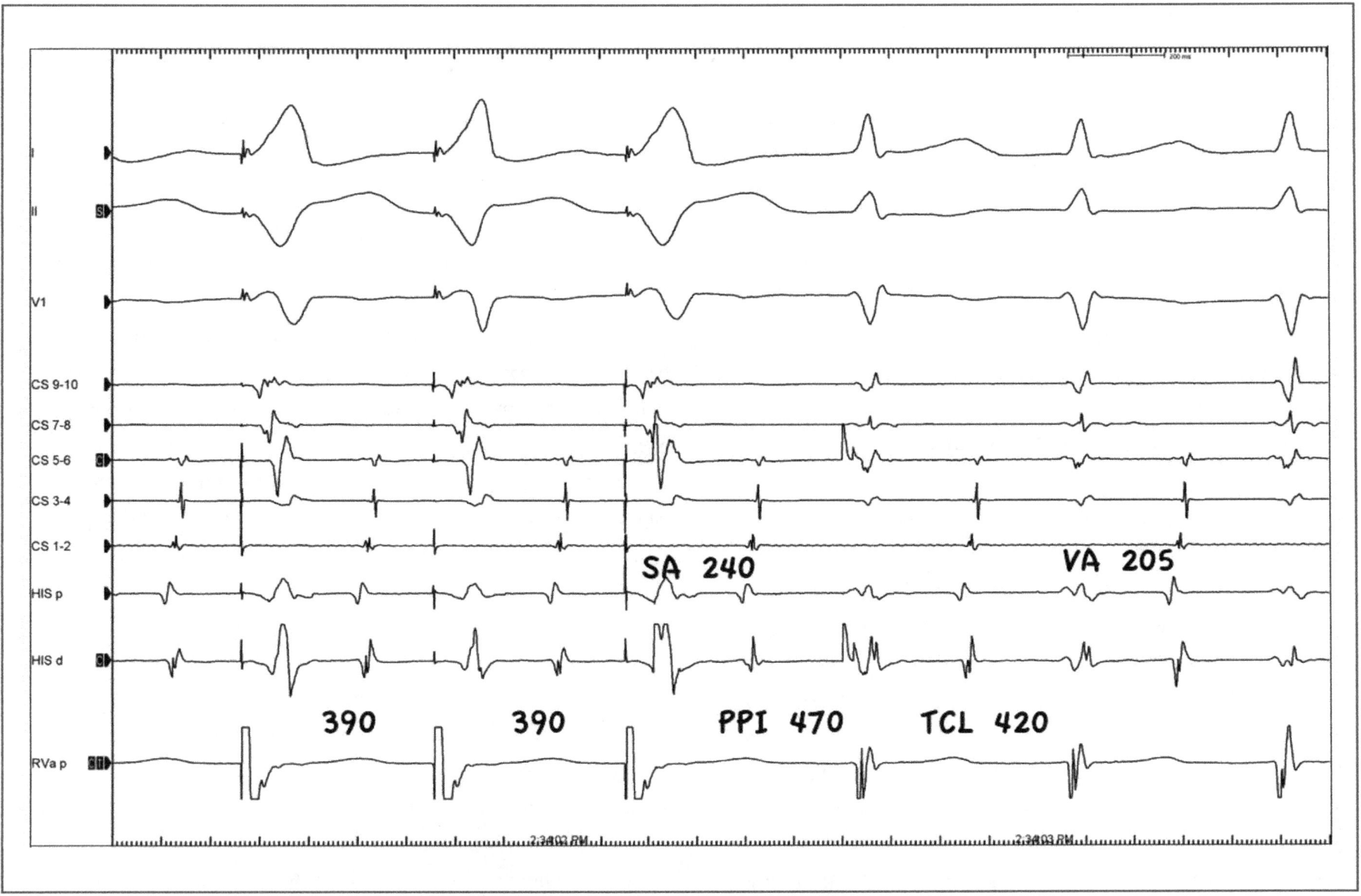

As a brief note, the initial values of PPI–TCL of 115 ms and SA-VA of 85 ms were originally described (selected reference 3) to differentiate orthodromic AVRT over septal APs from AVRT (less than 115 and 85 for septal AP and more than 115 and 85 for AVNRT). These values are useful more broadly and AP outside of the posteroseptal region often fall into this range. That is, the values can be more broadly thought of as indicating AVRT in general since values *less* than these are not seen in AVNRT. Values greater than these thresholds can be seen with some left lateral pathways in addition to AVNRT, since the RVA pacing site is relatively "far" from the circuit or, alternatively, when decremental conduction during entrainment prolongs conduction time over a component of the circuit.

**Figure 4.5** illustrates the utility of examining the tracing during the onset of ventricular overdrive pacing. In this example, the drive is begun during SVT with a septal atrial activation pattern. The first 2 stimuli don't perturb the tachycardia appreciably, but the third

results in obvious fusion of the paced QRS with the tachycardia QRS and this advances the next A. This is a typical example of "fusion with reset" and can be considered as essentially identical to advancement of the A by a His refractory PVC (see chapter 3). To reiterate the explanation, once conduction has exited the AV node, a paced beat from the ventricle cannot advance the next atrial EGM if the node is the only route of retrograde conduction, so there must be an AP involved in the circuit for all practical purposes.

**Figure 4.6** is very similar to Figure 4.5. In this example, AVRT terminates during ventricular pacing precluding measurement of the PPI, but the mechanism is evident at the onset of pacing by atrial reset (arrow) after minimal fusion evident in the prior QRS complex. This is, of course, not terribly unexpected since the mere fact of easy termination during relatively slow ventricular overdrive pacing suggests in itself that RV pacing has excellent access to the tachycardia circuit and is most compatible with AV reentry. It is difficult to imagine AVNRT or AT behaving in a similar way.

Figure 4.5

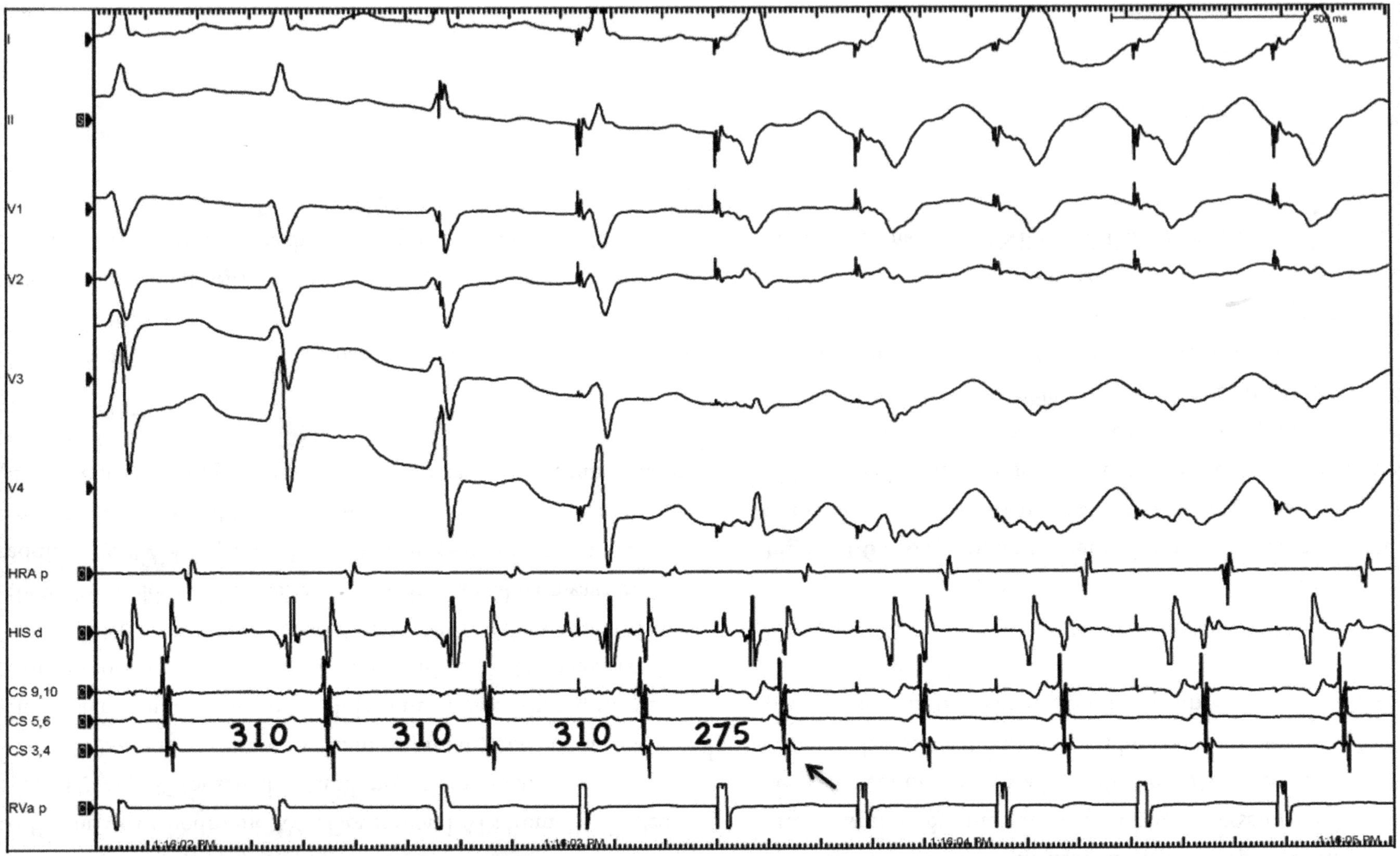

**Figure 4.6**

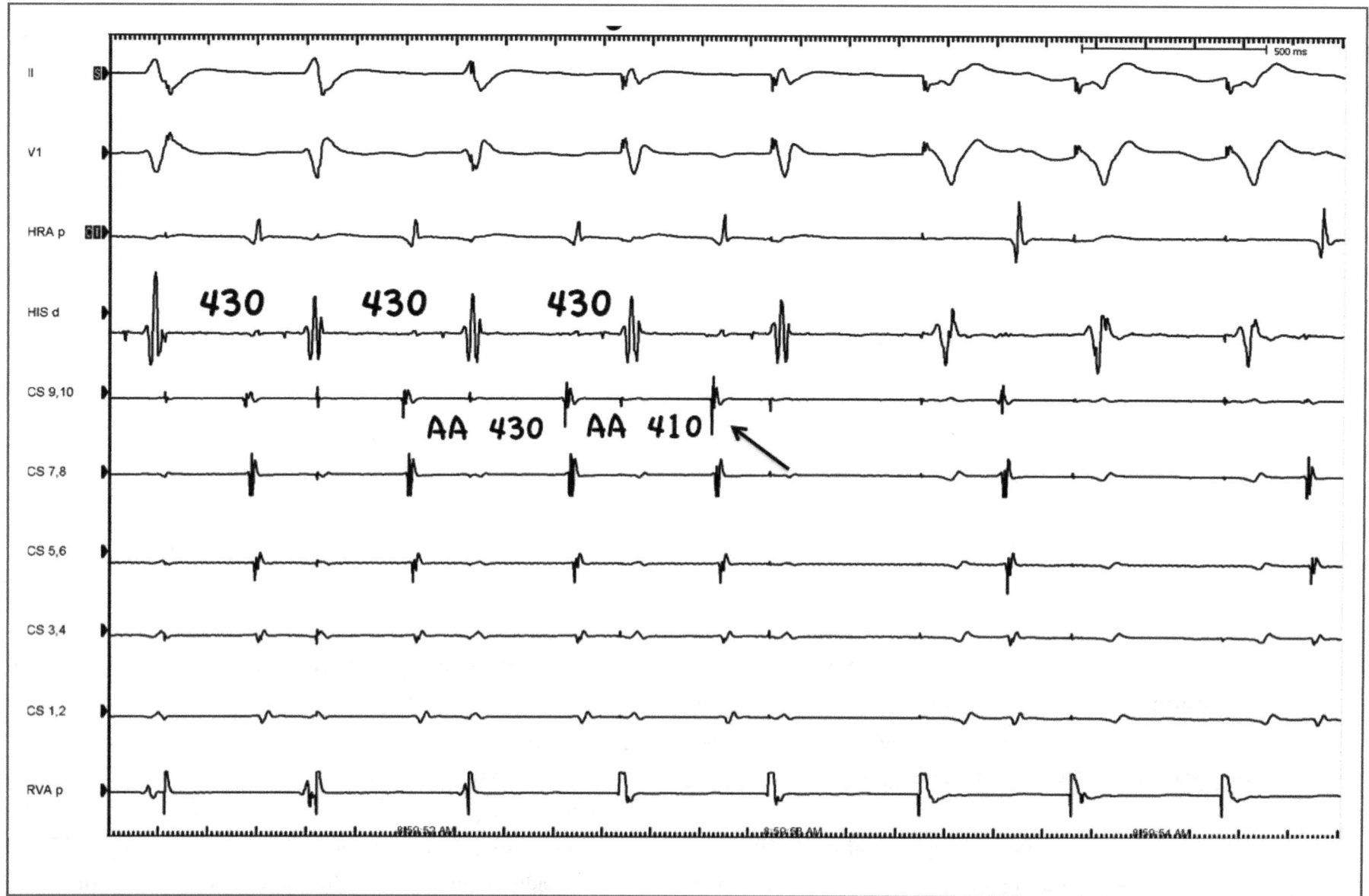

In contrast to the previous example, **Figure 4.7** is an example of AVNRT at the onset of overdrive pacing. Atrial reset (arrow) is only evident after at least 5 cycles of ventricular pacing and only when the QRS is completely paced (no fusion), and the His would not be expected to be refractory. This is, of course, essentially identical to failure to preexcite the next A during a His refractory PVC. It is useful to remember that preexciting the next A during His

refractoriness (i.e., reset and fusion) is proof of an AP but that *failure* to preexcite is not proof of its absence.

To the casual glance, **Figure 4.8** represents an "obvious" VAV response. However, closer inspection and measurement indicates that the AT has not accelerated to the pacing rate and there is no apparent VA conduction.

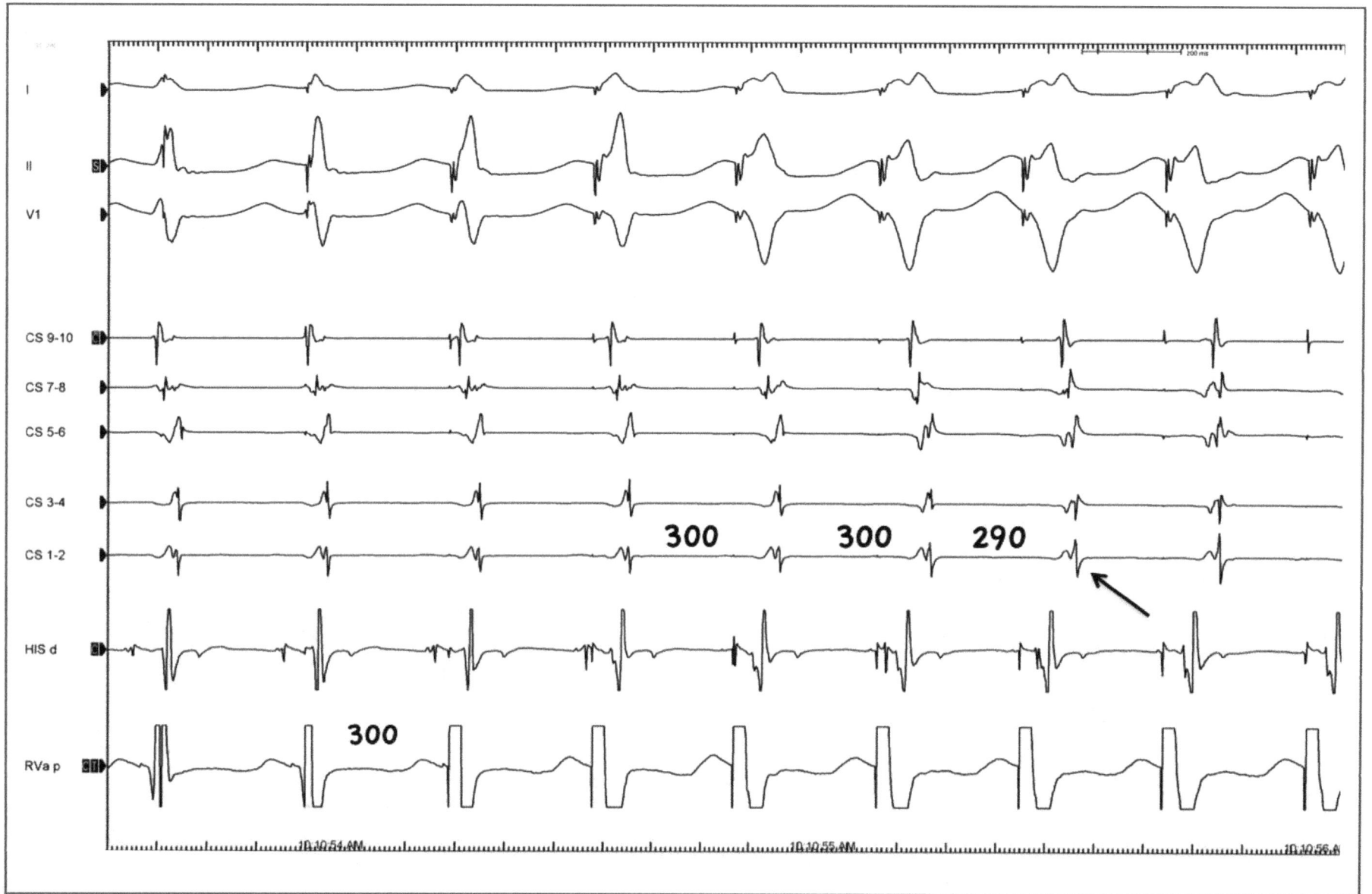

Figure 4.7

**Figure 4.8**

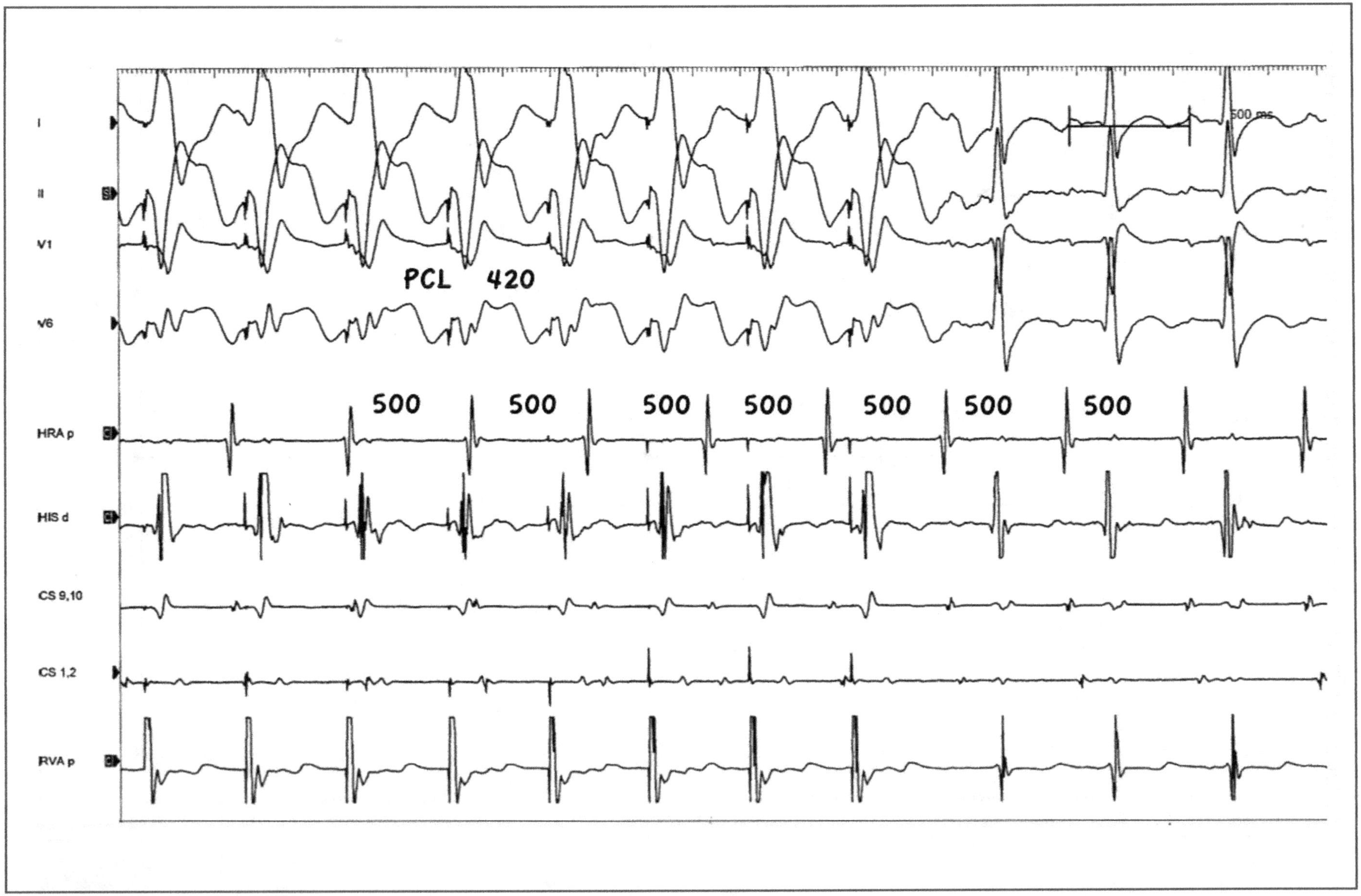

**Figure 4.9** shows ventricular overdrive pacing during SVT with a "short" AV interval. It is an apparent VAAV response. However, it must be remembered that the "A" in AVNRT can come before, during, or after the V since there is no true direct VA conduction during tachycardia. That is, the timing of the A and V are relative. In this example, the His EGM, although of poor amplitude, is recorded clearly before the second A, which can be thought of as comparable to a "retrograde" A even though it comes before the V. This is frequently referred to as a VAHV response but the only substantive issue is not to mistake this for a VAAV response.

**Figure 4.10** illustrates overdrive pacing of a regular, long VA (long RP) tachycardia that may be AT, AVNRT (atypical of course), or AVRT. Application of the checklist makes the mechanism quite clear. The overdrive pacing shows a "clean" acceleration of the tachycardia to the pacing rate. The last atrial cycle influenced by pacing is identified by measurement (short arrow) and is also identical in atrial activation sequence to that in tachycardia. This is

clearly a VAV response, excluding AT. Measurement of the PPI minus TCL quickly yields a value of 40 ms and subtraction of the VA from the SA during pacing gives 31 ms, both indicating close proximity to the reentrant circuit, most compatible with AV reentry utilizing a posteroseptal AP.

Note also, the His deflection (inset) during ventricular pacing and during tachycardia. The His during pacing has identical morphology to that during orthodromic tachycardia, indicating that the His is activated in an anterograde direction during ventricular pacing This is compatible with the collision point of the advancing AVN impulse and the retrograde impulse from ventricular pacing being distal to the recorded His bundle. The fact that the atrial EGM is accelerated during pacing indicates that an alternate retrograde route, the atrium (i.e., an AP), is being used since the pacing does not get to the AV node. This is "reset" with ventricular pacing while the His is refractory and is of identical significance to advancing the A by a His refractory PVC during tachycardia. This can only be AVRT.

**Figure 4.9**

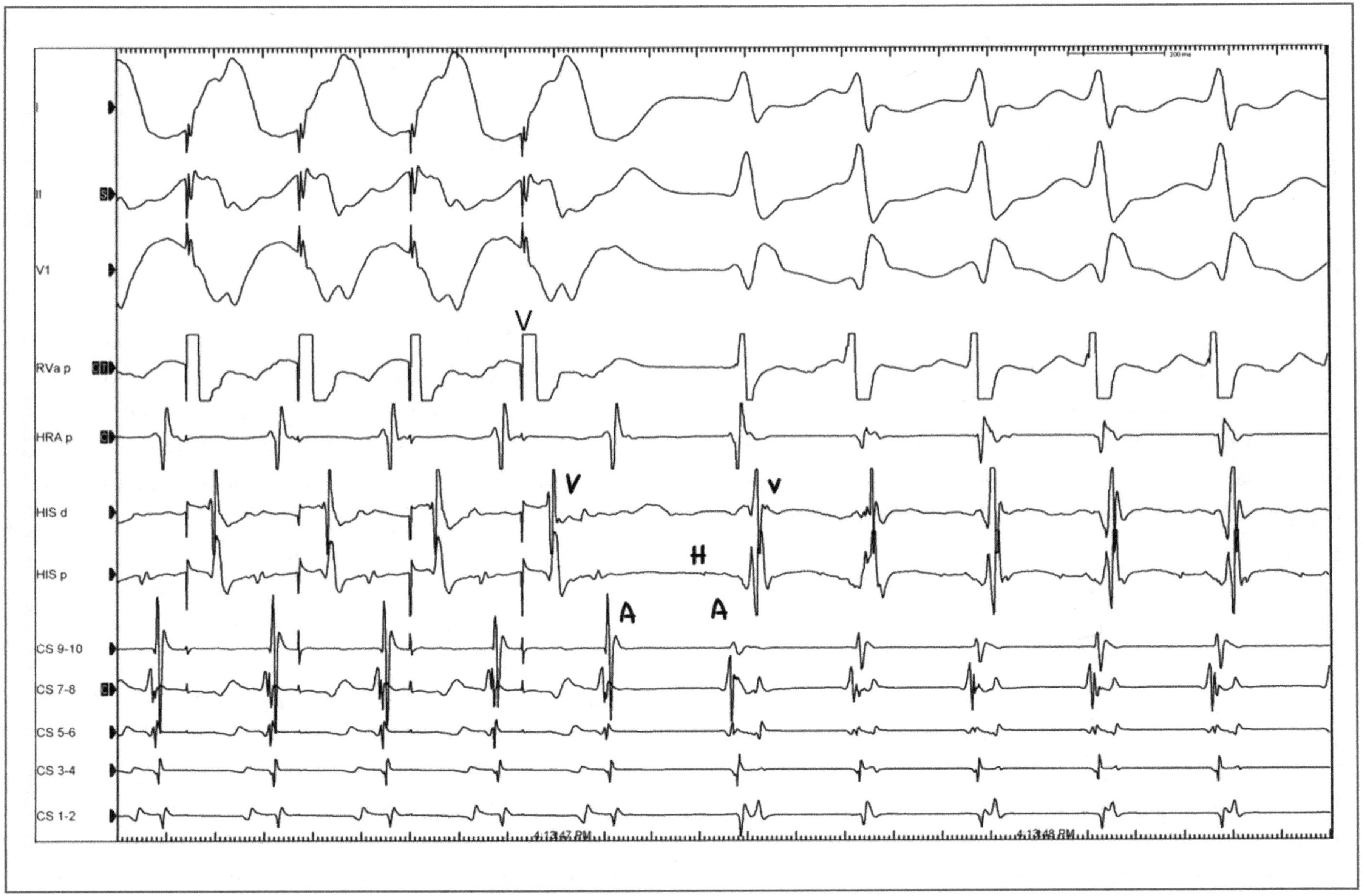
I
II
V1
RVa p
HRA p
HIS d
HIS p
CS 9-10
CS 7-8
CS 5-6
CS 3-4
CS 1-2
V
V
V
H
A
A
200 ms
4:13:47 PM
4:13:48 PM

**Figure 4.10**

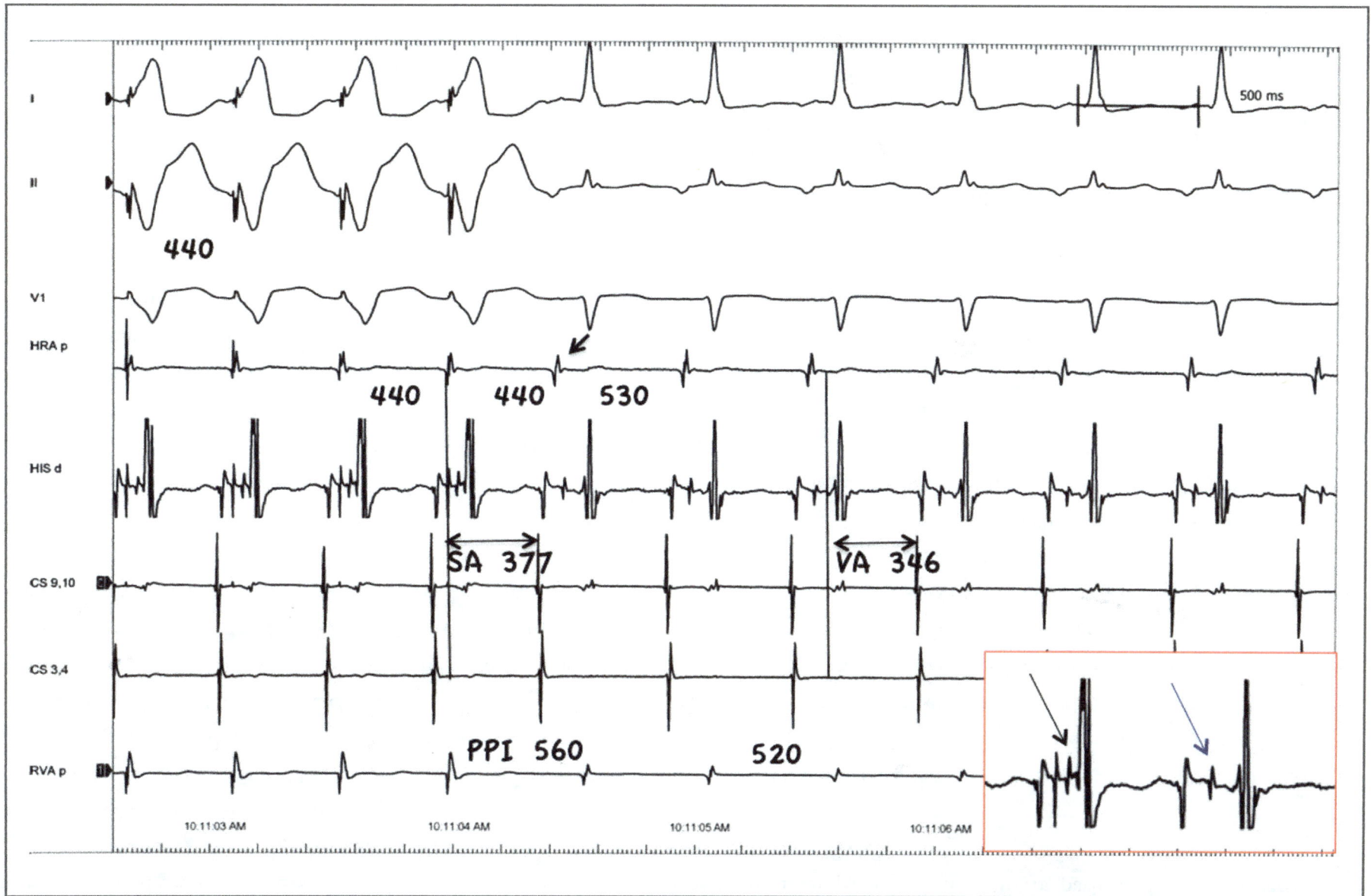

**Figure 4.11** at first glance indicates a VAAV response but the key to interpretation is application of the checklist and especially in identifying the last atrial cycle accelerated by pacing (indicated by short arrow) showing that it is a VAV response with a very long VA interval. It might be noted also that the atrial activation sequence during pacing is identical to that of tachycardia, more compatible with entrainment of AVRT or AVNRT rather than overdrive of an AT. Also, one might note that the SA interval during entrainment cannot be *shorter* than the VA during tachycardia, hence ruling out the shorter SA interval possibility that allows the erroneous designation of VAAV response.

A subtlety emerges in designating this as due to AVNRT on the basis of the measured PPI minus TCL and SA-VA during tachycardia. These intervals are very long and more compatible with AVNRT as a rule. However, with long RP tachycardia related to AVRT over a slowly conducting AP, the measured intervals may well overlap with those of AVNRT and this is related to prolongation of conduction time over these decremental pathways during entrainment. Thus, this example is clearly not AT but other maneuvers would have to distinguish atypical AVNRT from AVRT using a decremental AP. One of these might be to compare entrainment from the ventricular apex and the base of the heart near the pathway at the same cycle lengths. One would then expect shorter intervals from basal entrainment in the case of an AP with

the entrainment pacing site closer to the pathway and the converse with atypical AVNRT.

**Figure 4.12** shows a regular wide QRS tachycardia with a 1:1 A to V relationship. A His deflection is clearly seen at the onset of the QRS and this can realistically only be VT or preexcited tachycardia. Earliest atrial activation is at the CS orifice, compatible with a "septal" activation pattern. This example shows that the entrainment checklist can be used in this type of example exactly as for orthodromic SVT. Note that the atrial activation sequence is identical during pacing and during tachycardia. The pacing rate is only marginally accelerated over the TCL and the QRS during pacing is clearly fused (as verified by examining the pure "paced" QRS and comparing to the tachycardia QRS). We identify the last atrial cycle at the paced cycle length (short arrow) and can describe a VAV response. The measured intervals (PPI minus TCL, SA minus VA) are relatively short, suggesting that the RV apical pacing site is "in" or relatively close to the circuit. This would not be compatible with AT or AVNRT using a bystander AP, and more compatible with a macroreentrant preexcited tachycardia, such as antidromic tachycardia. Reentrant VT coming from the RV can't be entirely ruled but one might observe that the last advanced A also advanced the first postpacing QRS with the identical AV interval as during tachycardia, that is, the V was "linked" to the retrograde A. This was indeed antidromic AVRT using an atriofascicular pathway as the anterograde limb of the circuit.

**Figure 4.11**

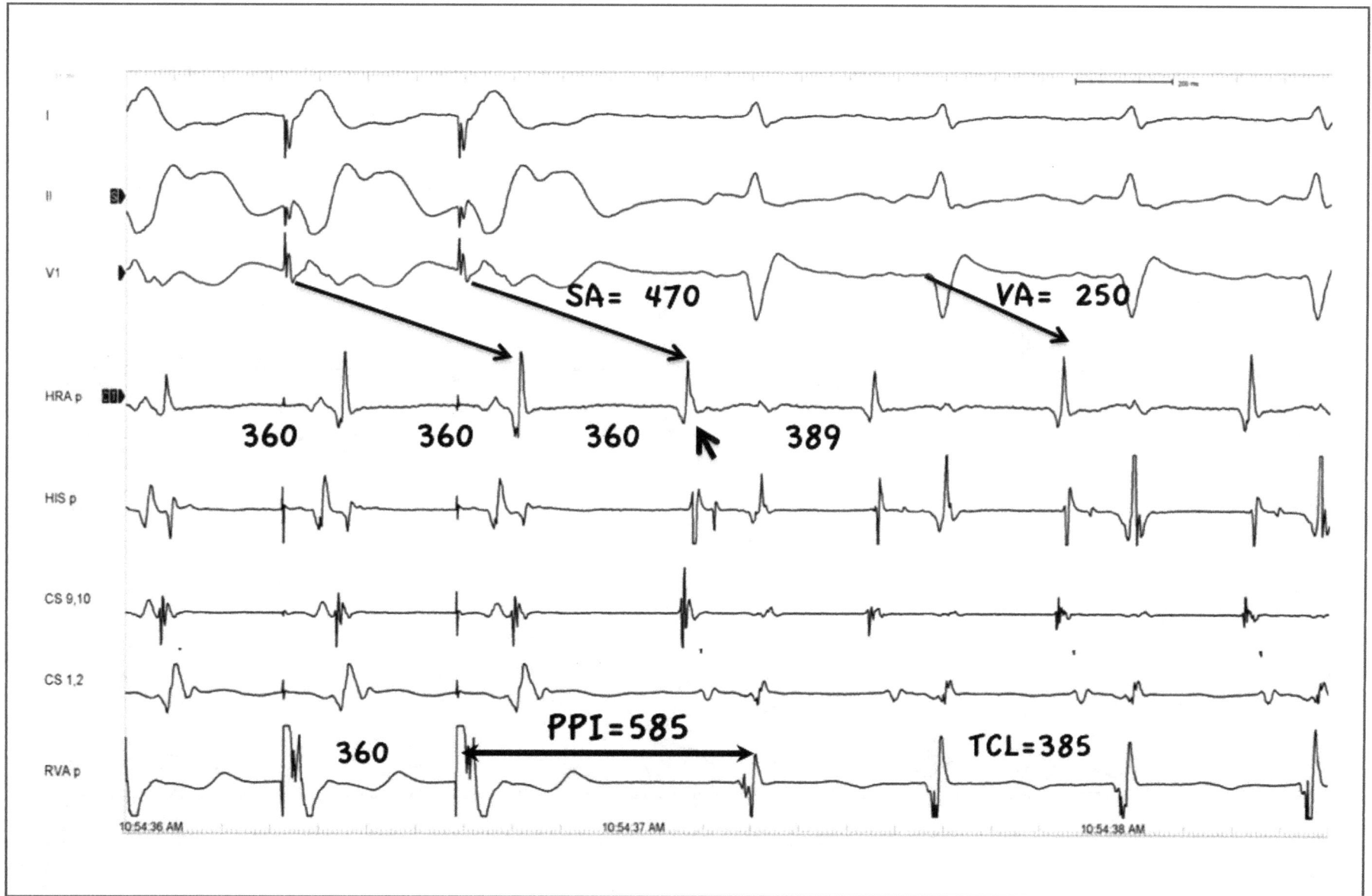

Figure 4.12

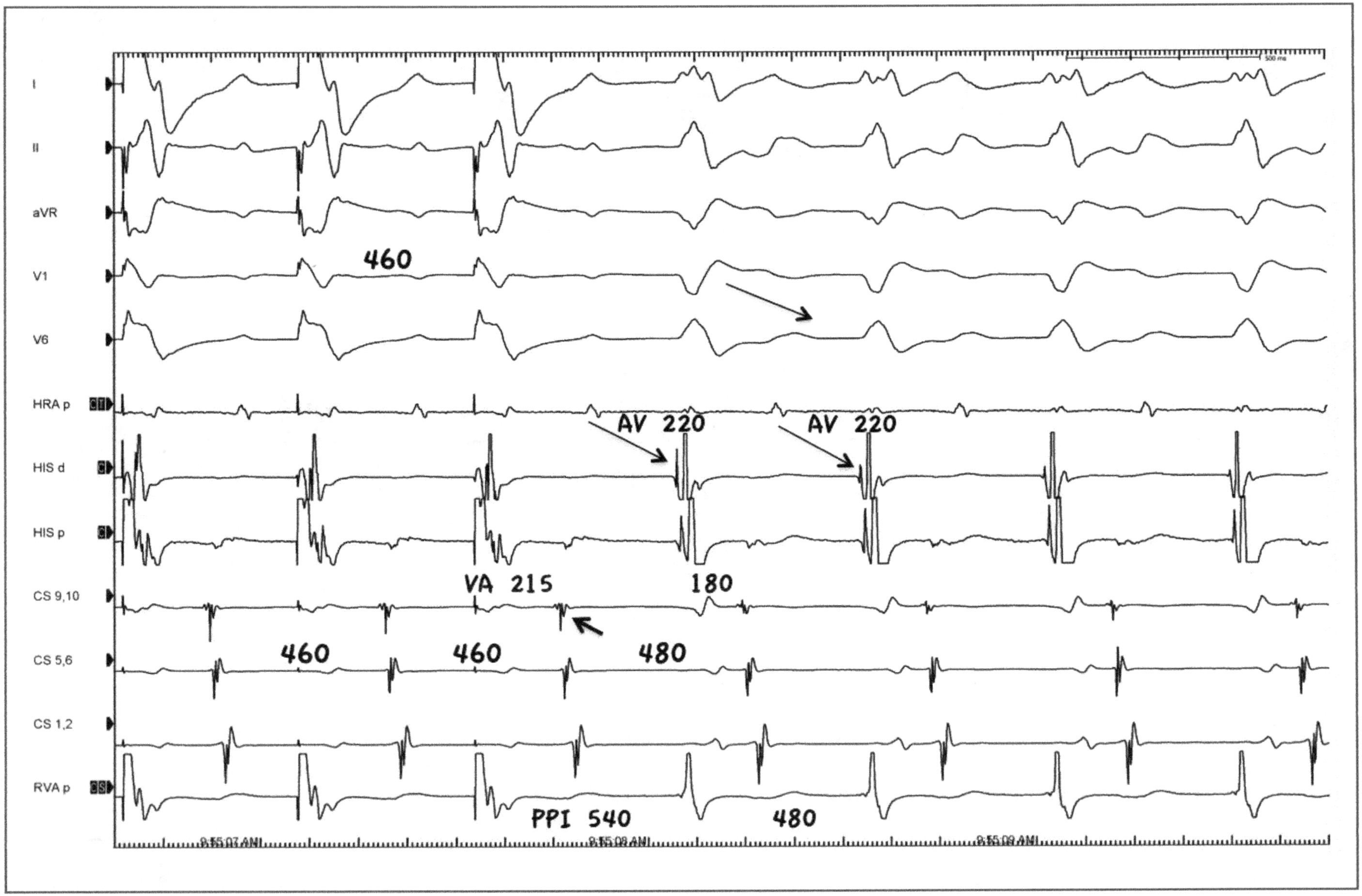
I
II
aVR
V1
V6
HRA p
HIS d
HIS p
CS 9,10
CS 5,6
CS 1,2
RVA p
460
AV 220
AV 220
VA 215
180
460
460
480
PPI 540
480
500 ms
9:55:07 AM
9:55:08 AM
9:55:09 AM

Atrial entrainment is generally less useful during tachycardia assessment but, not surprisingly, utilizes identical principles and is also facilitated in interpretation by a checklist. In the relatively simple example shown to illustrate the principles in **Figure 4.13**, a wide QRS tachycardia is seen with one to one, A to V relationship. The differential diagnosis is quite broad from the limited leads provided. Pacing is begun during tachycardia and the cycle length of the tachycardia is accelerated, maintaining the same QRS morphology. This obviously makes VT very unlikely but this can be verified further. The last ventricular cycle accelerated to the pacing is identified by measuring intervals (arrow). The next event is an atrial event that follows the advanced QRS. One might call this an A-V-A response. If this were VT, one would expect the first event after the overdriven QRS to be a ventricular event, that is, an A-V-V-A response. This was a preexcited antidromic tachycardia.

**Figure 4.14** highlights a use of entrainment during catheter ablation. In this patient with a unidirectional AP, AP conduction was not clearly evident during ventricular pacing due to continuing AVN conduction. On the other hand, ablation during SVT resulted in catheter instability at the termination of tachycardia. Entrainment allowed mapping of the atrial insertion of the AP during ablation with continued VA conduction after loss of AP conduction. The first 4 cycles during entrainment show earliest retrograde atrial activation at the distal CS (CS3, 4) and the RF abl catheter is at the target site on the AV ring with shortest obtainable VA conduction. The AP blocks with the fifth cycle and no VA conduction is noted for one cycle as a result of concealed anterograde conduction into the AV node from AP conduction of the preceding beat. The following cycle shows a slightly altered atrial activation perhaps due to atrial ectopy but thereafter VA conduction occurs over the normal VA conduction system (CS 9,10 earliest) and the catheter remains stable while ablation is completed.

Figure 4.13

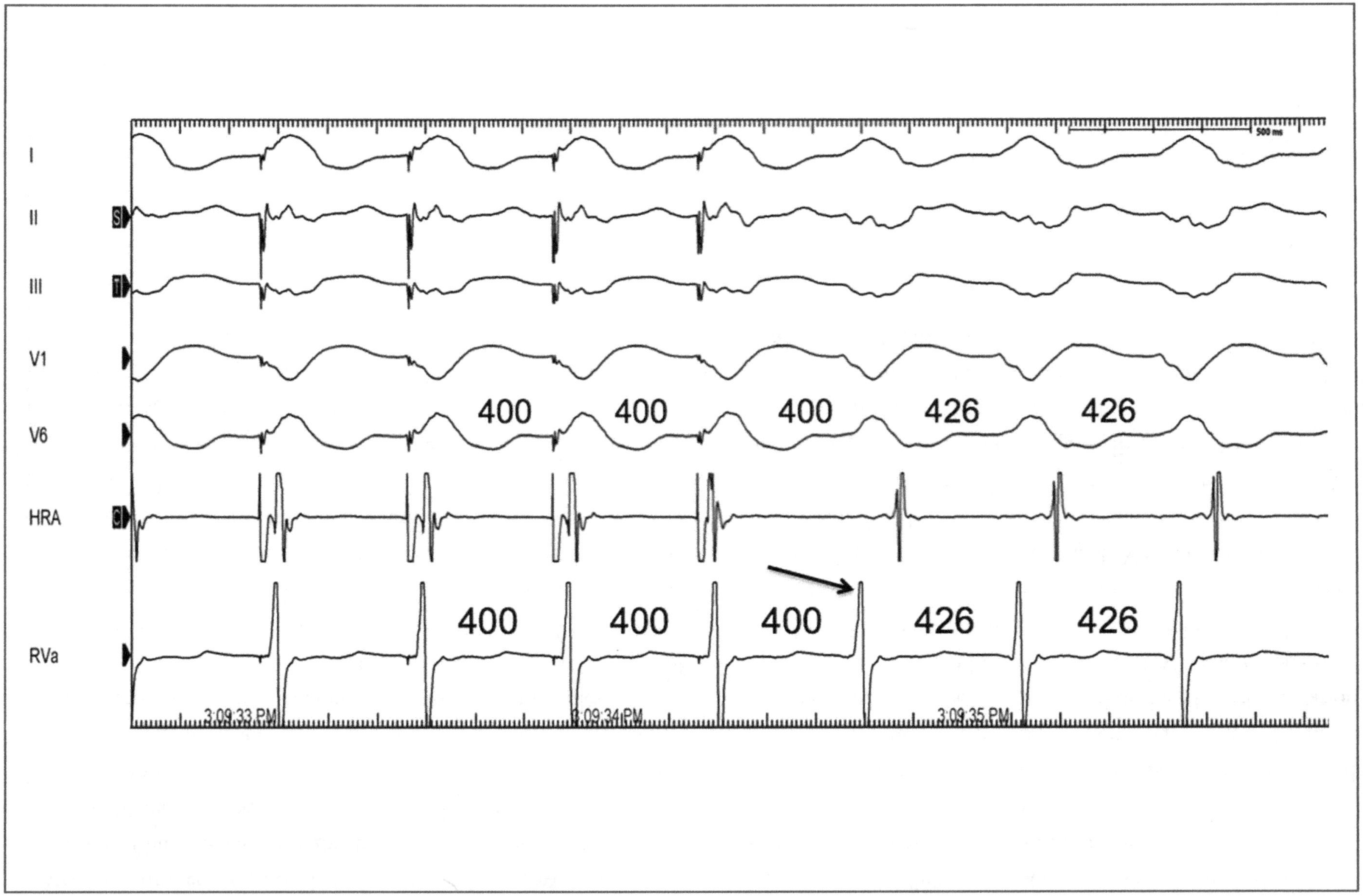

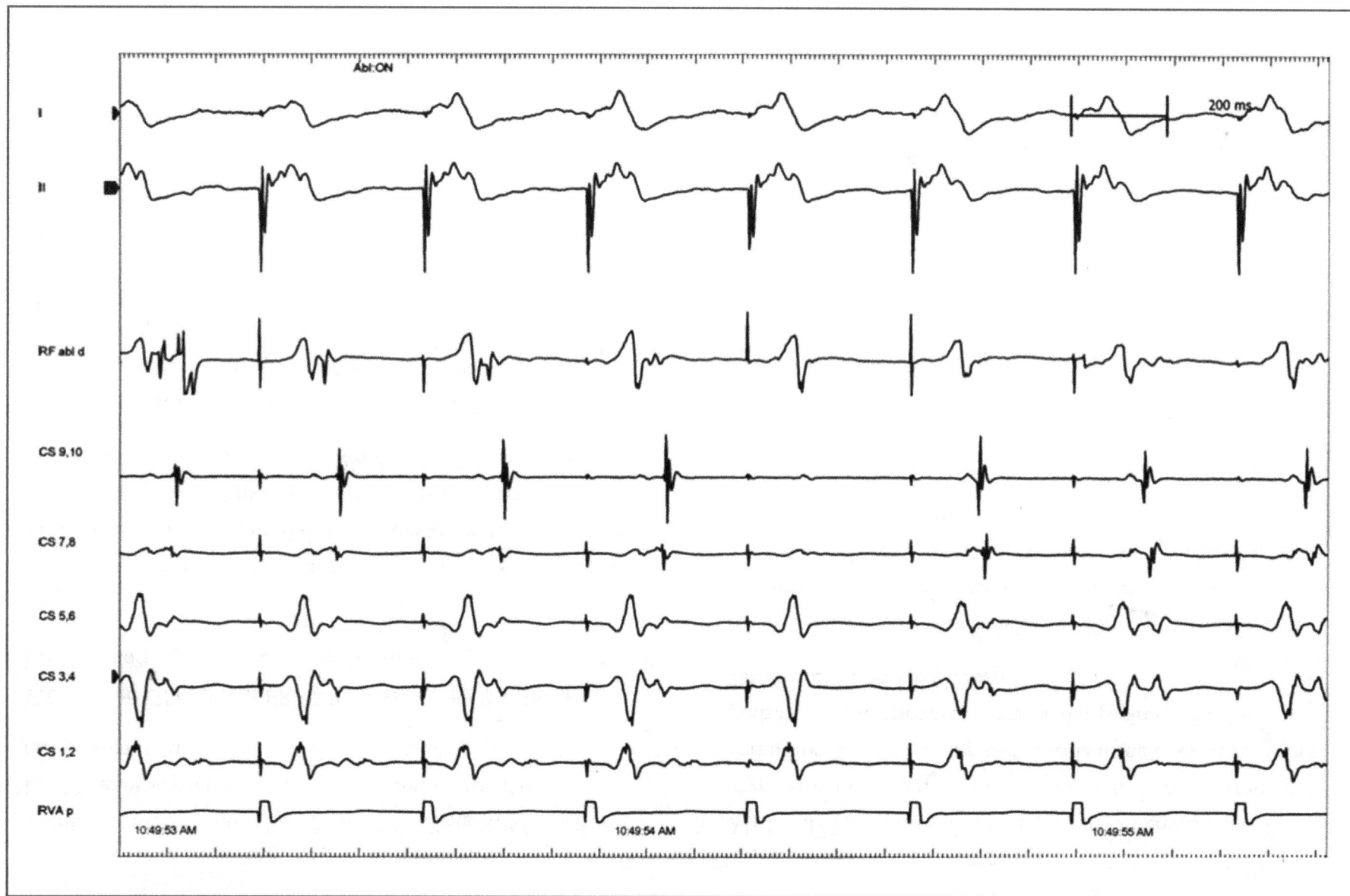

Figure 4.14

## Suggested Readings

1. Veenhuyzen GD, Quinn FR, Wilton SB, Clegg R, Mitchell LB. Diagnostic pacing maneuvers for supraventricular tachycardia: Part 1. *PACE.* 2011;34:767–782.

2. Veenhuyzen GD, Quinn FR, Wilton SB, Clegg R, Mitchell LB. Diagnostic pacing maneuvers for supraventricular tachycardia: Part 2. *PACE.* 2012;35:757–769.

3. Obeyesekere M, Gula LJ, Modi S, et al. Tachycardia induction with ventricular extrastimuli differentiates atypical atrioventricular nodal reentrant tachycardia from orthodromic reciprocating tachycardia. *Heart Rhythm.* 2012;9:335–341.

4. Abdelwahab A, Gardner MJ, Basta MN, et al. A technique for the rapid diagnosis of wide complex tachycardia with 1:1 AV relationship in the Electrophysiology Laboratory. *PACE.* 2009;32:475–483.

5. Michaud GF, Tada H, Chough S, et al. Differentiation of atypical atrioventricular node re-entrant tachycardia from orthodromic reciprocating tachycardia using a septal accessory pathway by the response to ventricular pacing. *J Am Coll Cardiol.* 2001;38:1163–1167.

6. Segal OR, Gula LJ, Skanes AC, et al. Differential ventricular entrainment: A maneuver to differentiate AV node reentrant tachycardia from orthodromic reciprocating tachycardia. *Heart Rhythm.* 2009;6:493–500.

# Overdrive Pacing and Entrainment:

## Ventricular Tachycardia, Atrial Tachycardia, and Flutter

The diagnosis of tachycardia mechanism is generally obvious after a full diagnostic study and the performance of the maneuvers described in the preceding chapters. The next challenge is the ablation of the tachycardia. In the cases of atrial tachycardia

(AT) and ventricular tachycardia (VT), this is largely dependent upon activation or entrainment mapping, or alternatively, substrate-guided ablation when the former is not possible. From a practical point of view, the tachycardia mechanism is classified either as "focal" or macroreentrant. It is appreciated that "focal" does not really distinguish an automatic or triggered focus from a micro-reentrant circuit, but these distinctions may be difficult to make in the laboratory and not necessary for the practical consideration of mapping and ablation. The "focal" tachycardias, AT or VT, involve activation emanating from a relative point source that is the target of ablation. Complexities can arise when tachycardias are difficult to induce or when more than one mechanism is at play, but the fundamental task of tracing activation to a point source remains straightforward.

The macroreentrant tachycardias are more challenging in that not only does the circuit have to be defined, but also critical areas where ablation will prevent the tachycardia from circulating. These are often in a "slow conduction" zone where electrograms (EGMs) may be of low voltage or fragmented, and the functional significance of the EGMs recorded is not readily apparent. Entrainment mapping is very useful in defining whether a given site is close to the circuit (i.e., "in" or "out" of the circuit) and also can provide clues as to whether it is an important or obligatory part of the circuit. The concept of critical parts of the circuit becomes clear when one considers that typical atrial flutter involves a large right atrial circuit where many sites are "in" (probably much of the right atrium) but the usual target of ablation is the tricuspid-caval isthmus, which is the narrowest part of the circuit and most practical to ablate.

# Ventricular Tachycardia

**Figure 5.1** illustrates VT at cycle length (CL) of 300 ms suspected to be bundle branch reentry. The pacing catheter is positioned near the right bundle branch (RBB) position and tachycardia is entrained at CL of 280 ms (progressive fusion demonstrated elsewhere). The ECG during pacing closely resembles that of tachycardia but is manifestly fused due to activation of both the RBB and adjacent muscle (fully paced QRS not illustrated). The postpacing interval (PPI) is virtually identical to the tachycardia CL showing that the pacing site is "in" the circuit and supports the mechanism of bundle branch reentry. Tachycardia could no longer be induced after ablation of the RBB. (Tracing compliments of Dr. R. Sy.)

**Figure 5.2** is taken from a patient with a presumed focal "slow" left ventricular (LV) tachycardia. Pacing the posteroseptal LV near the left posterior fascicle overdrives tachycardia, which resumes on termination of pacing. The PPI is 60 ms, suggesting that the pacing site is not in the circuit or at the focal source. **Figure 5.3** in the same patient illustrates the same maneuver pacing more anteriorly, at a site near the left anterior fascicle where a presystolic potential (arrow) is recorded. This reproduces the VT morphology with a short stimulus to QRS and suggests close proximity to the exit site of a circuit if reentrant or, alternatively, a focus. The PPI is only 10 ms at this site and this was the successful ablation site.

Whereas the above examples relate to VT without scar, much monomorphic VT occurs in the context of ischemic or myopathic scar that provide the substrates for reentry. This substrate is invariably associated with low-voltage areas and delayed, fractionated potentials that may or may not be important to a tachycardia mechanism. The latter can be used to guide ablation if tachycardia is not amenable to mapping, but mapping clearly provides the best opportunity to ablate the tachycardia mechanism specifically. Entrainment is probably the best tool to facilitate this and pacing is performed just fast enough to be certain of acceleration, but not too fast to avoid pacing-induced conduction delay that confounds results.

When pacing is terminated after VT is entrained, the return cycle at the pacing site reflects the time it takes to travel to the circuit, around the circuit, and back to the pacing site (the PPI). This should approximate the tachycardia cycle length (TCL) if the site is in the circuit (practically speaking, PPI–TCL < 30 ms); PPI–TCL progressively lengthens as the pacing site is moved farther from the circuit. *Note that this does not necessarily represent a "geographical" distance as a line of block or conduction delay may make a pacing site seem far from the circuit "functionally," but relatively close "geographically."*

Figure 5.1

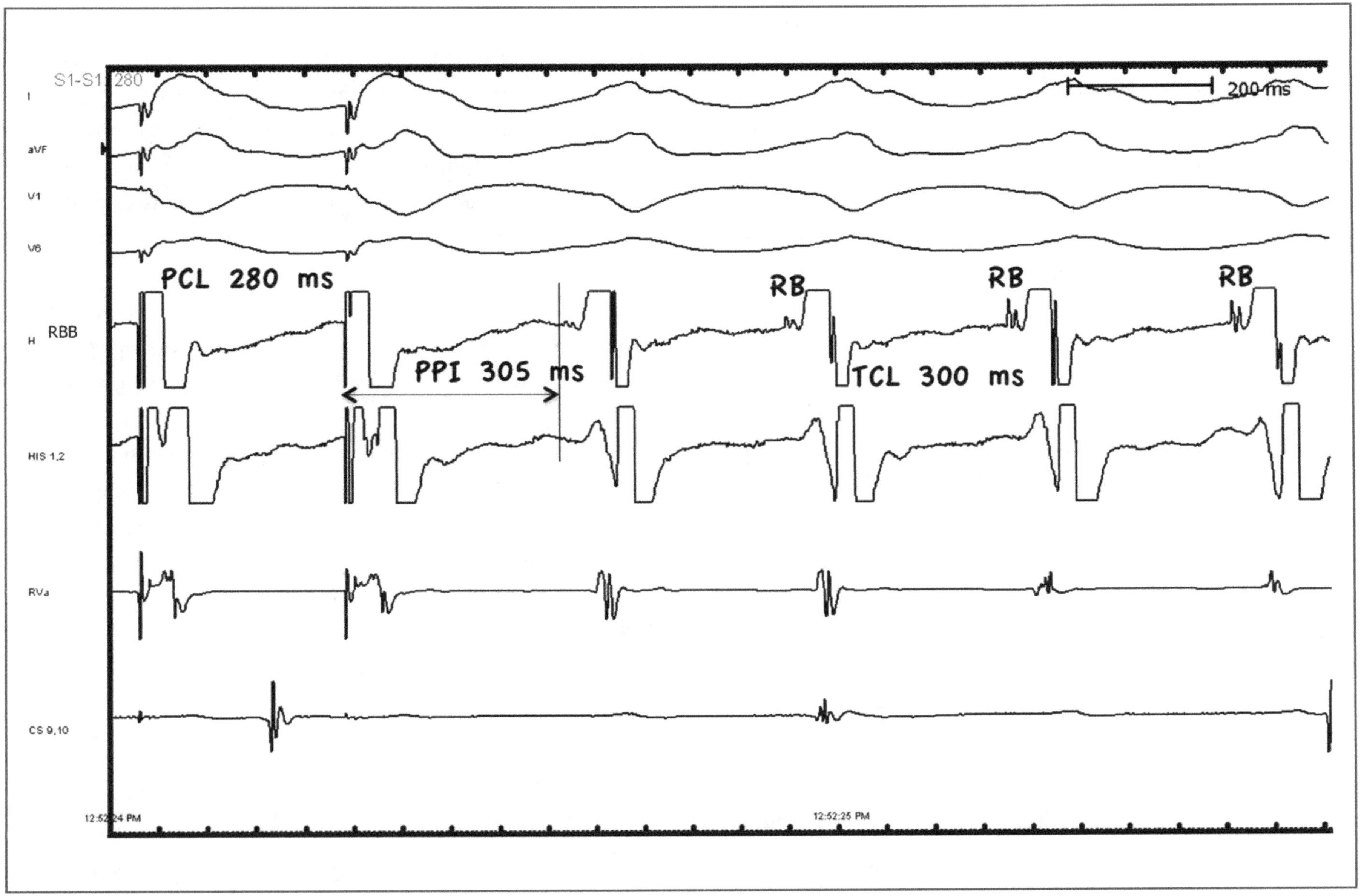
S1-S1 280
200 ms
I
aVF
V1
V6
PCL 280 ms
RB
RB
RB
H RBB
PPI 305 ms
TCL 300 ms
HIS 1,2
RVa
CS 9,10
12:52:24 PM
12:52:25 PM

**Figure 5.2**

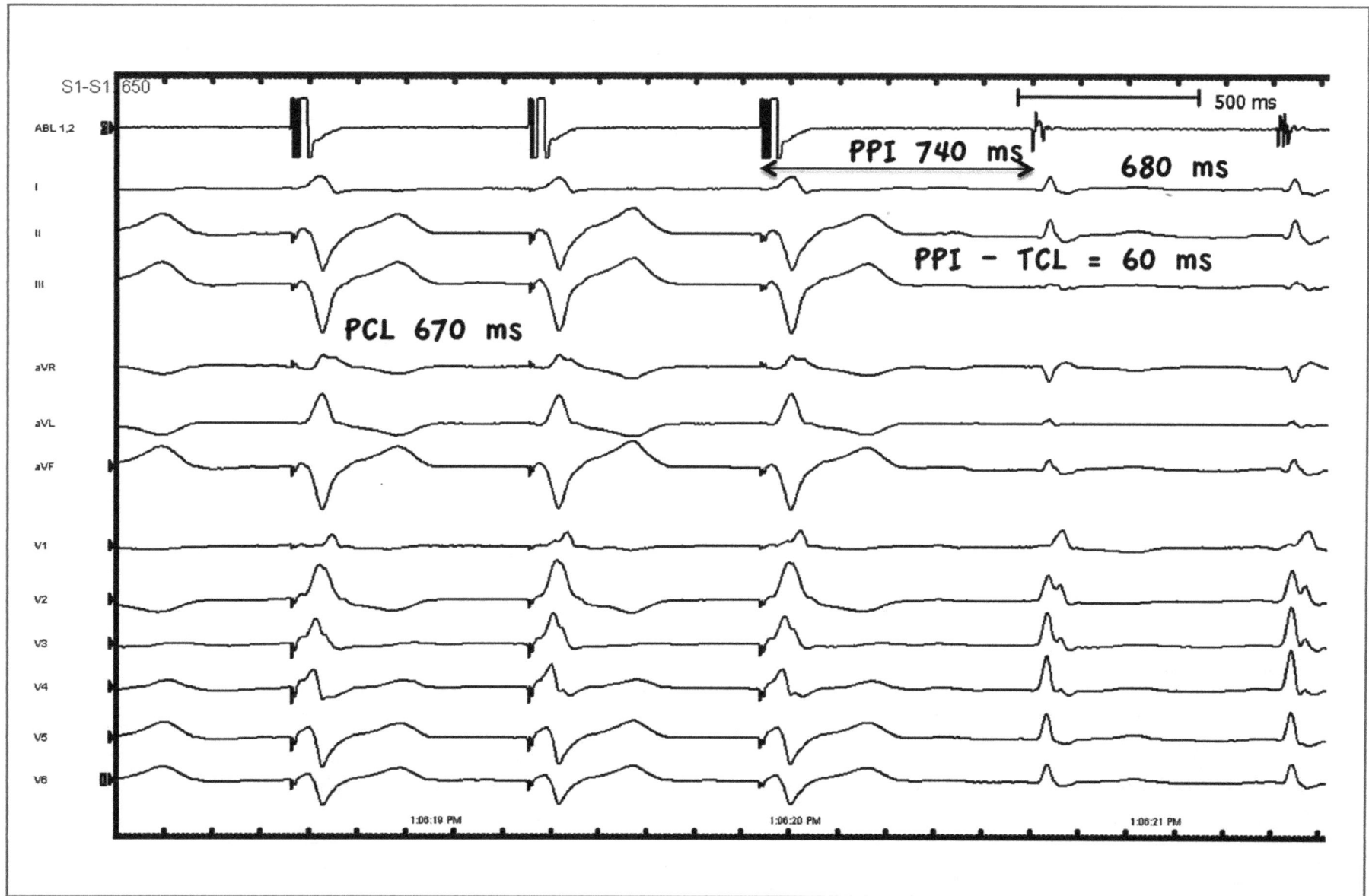

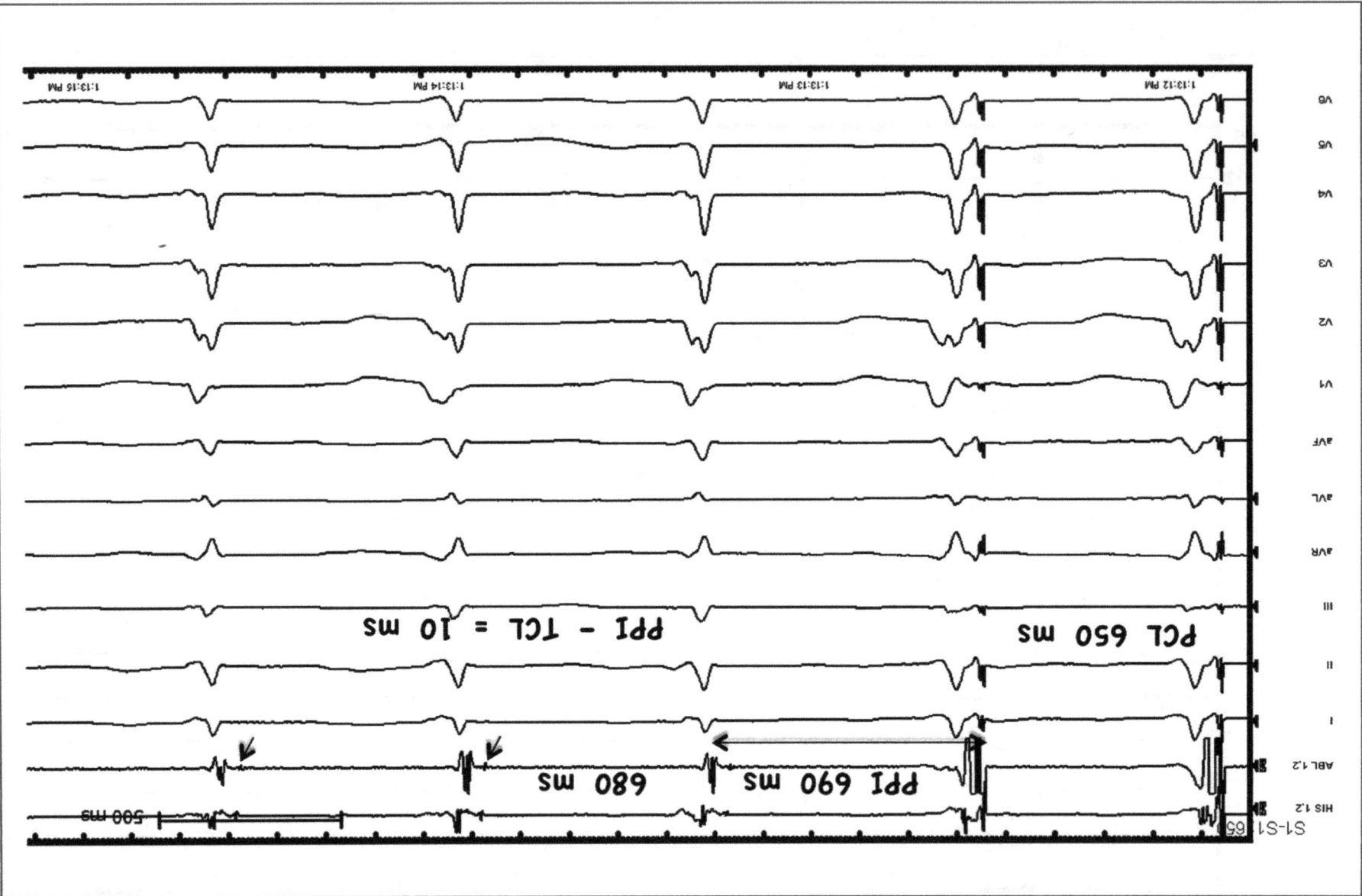

**Figure 5.3**

The extent of delay from the stimulus to the QRS (S-QRS) can further characterize a catheter's relative position to the circuit within a scar, with a longer S-QRS representing a pacing site that is either in a blind loop of the scar or in a proximal position of the circuit relative to the exit of the scar.

Finally, the presence of manifest fusion of the QRS reflects the presence of both an orthodromic and antidromic wave. Such a site may be "in" the circuit but not in the critical slow-conduction zone of the scar. When pacing in a protected area of the scar, fusion is undoubtedly occurring to some degree within the scar but it is not electrocardiographically apparent (entrainment with *concealed* fusion).

This is due to relative unidirectional block behind the advancing wave resulting from pacing.

**Figures 5.4** and **5.5** are based on the Suggested Readings. Together they provide a reasonable working hypothetical schema to assess the relative position of a given recording site within a VT substrate and its possible candidacy as an ablation site. The subsequent figures will include representative tracings showing how entrainment data fit into the schema of Figures 5.4 and 5.5. It will be useful to refer back to this schema as these subsequent examples are discussed.

A checklist as a starting point to these tracings might be:

1. Is the tachycardia properly entrained?

2. Is there manifest or concealed fusion?

3. Is the PPI "in"?

4. What is the S-QRS relative to the EGM-QRS?

The interested reader is strongly urged to read the classical publication by Stevenson et al (see Suggested Readings) and others at the end of this chapter for a more complete discussion.

**Figure 5.4**

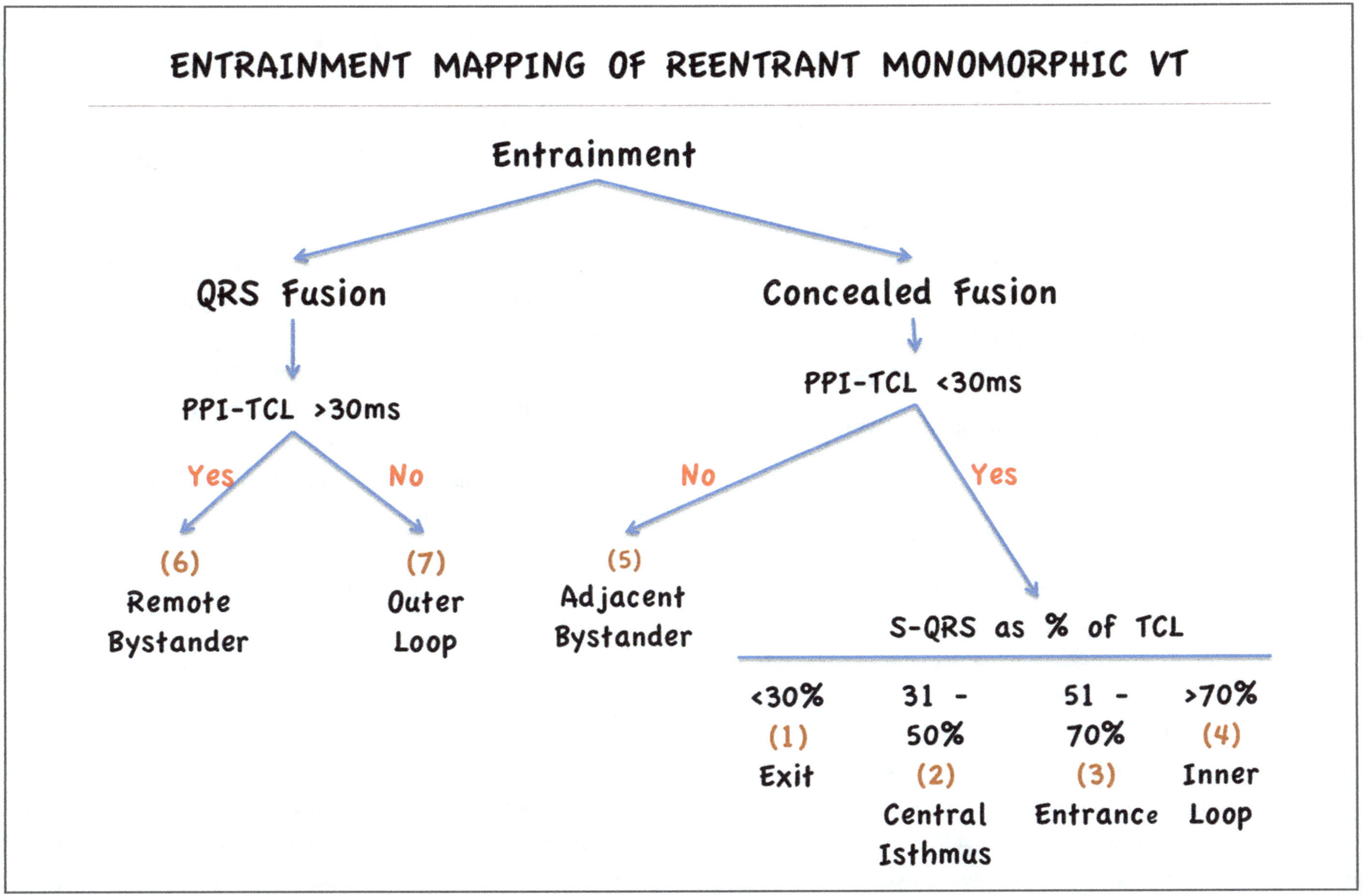

# ENTRAINMENT MAPPING OF REENTRANT MONOMORPHIC VT

Entrainment

QRS Fusion

Concealed Fusion

PPI-TCL >30ms

PPI-TCL <30ms

Yes       No

No       Yes

(6)       (7)       (5)

Remote       Outer       Adjacent
Bystander       Loop       Bystander

S-QRS as % of TCL

| <30% | 31 – 50% | 51 – 70% | >70% |
|------|----------|----------|------|
| (1) | (2) | (3) | (4) |
| Exit | Central Isthmus | Entrance | Inner Loop |

**Figure 5.5**

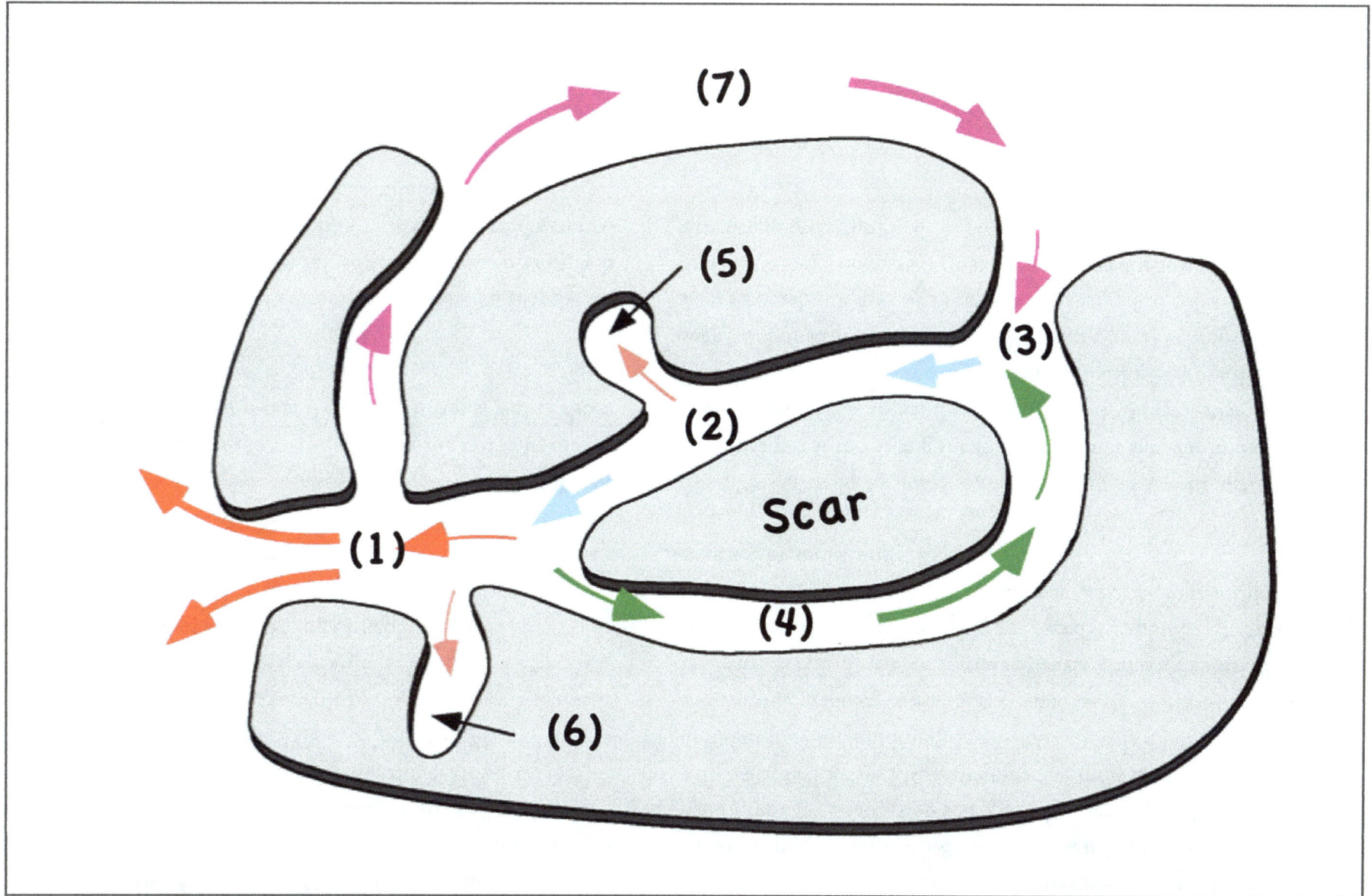

The issue in **Figure 5.6** is very fundamental but needs to be emphasized and questioned each time the entrainment is performed. The CL of pacing should be as close as possible to the TCL to avoid conduction delay in the circuit and this was done. However, the closer the CL of pacing is to the TCL, the longer it will take to entrain the circuit. In this instance, the pacing was not maintained long enough to catch up to the tachycardia as is clearly evident from inspection of the EGMs at the pacing catheter (arrows).

Overdrive pacing at the site in **Figure 5.7** entrains the tachycardia with a PPI-TCL of 10 ms, suggesting the pacing site to be "in" the circuit. However, the QRS is manifestly fused (pure-pacing QRS not shown). The preceding suggests that the pacing site is within the circuit but at an outer loop outside the critical isthmus within the scar (#7 in the schema of Figures 5.4 and 5.5).

To illustrate the link to classical entrainment, it is noted that the last-paced cycle (F, E) is "fused and entrained," whereas the next cycle is the last entrained (E) but not fused. It is to be remembered that each pacing stimulus of a fused QRS during entrainment is actually advancing the *subsequent* QRS and not the one it is fused with.

A pacing site that reproduces the QRS morphology of VT during entrainment as in **Figure 5.8** must be advancing the circuit from within the slow-conduction zone. This has been termed "entrainment with concealed fusion." Reset with pacing must involve some degree of fusion from an antidromic wave although this is not apparent electrocardiographically, and hence "concealed." The last-paced cycle is labeled as F, E (fused and entrained), even though the fusion is concealed, whereas the first cycle not paced is entrained, E, (sped up to the pacing cycle length (PCL)) but not fused. Note that this is achieved by measuring CL on the surface ECG.

The PPI is virtually identical to the TCL (PPI-TCL = 10 ms), confirming that the pacing site is at an isthmus site within the circuit. The delay between the stimulus to the QRS is ~50% of the TCL and approximates the delay from the local EGM to QRS, suggesting that this pacing site is likely in a central isthmus site (site 2 on schema). It is assumed that this is a theoretical construct only and that a long conduction time does not necessarily represent a long distance geographically.

**Figure 5.6**

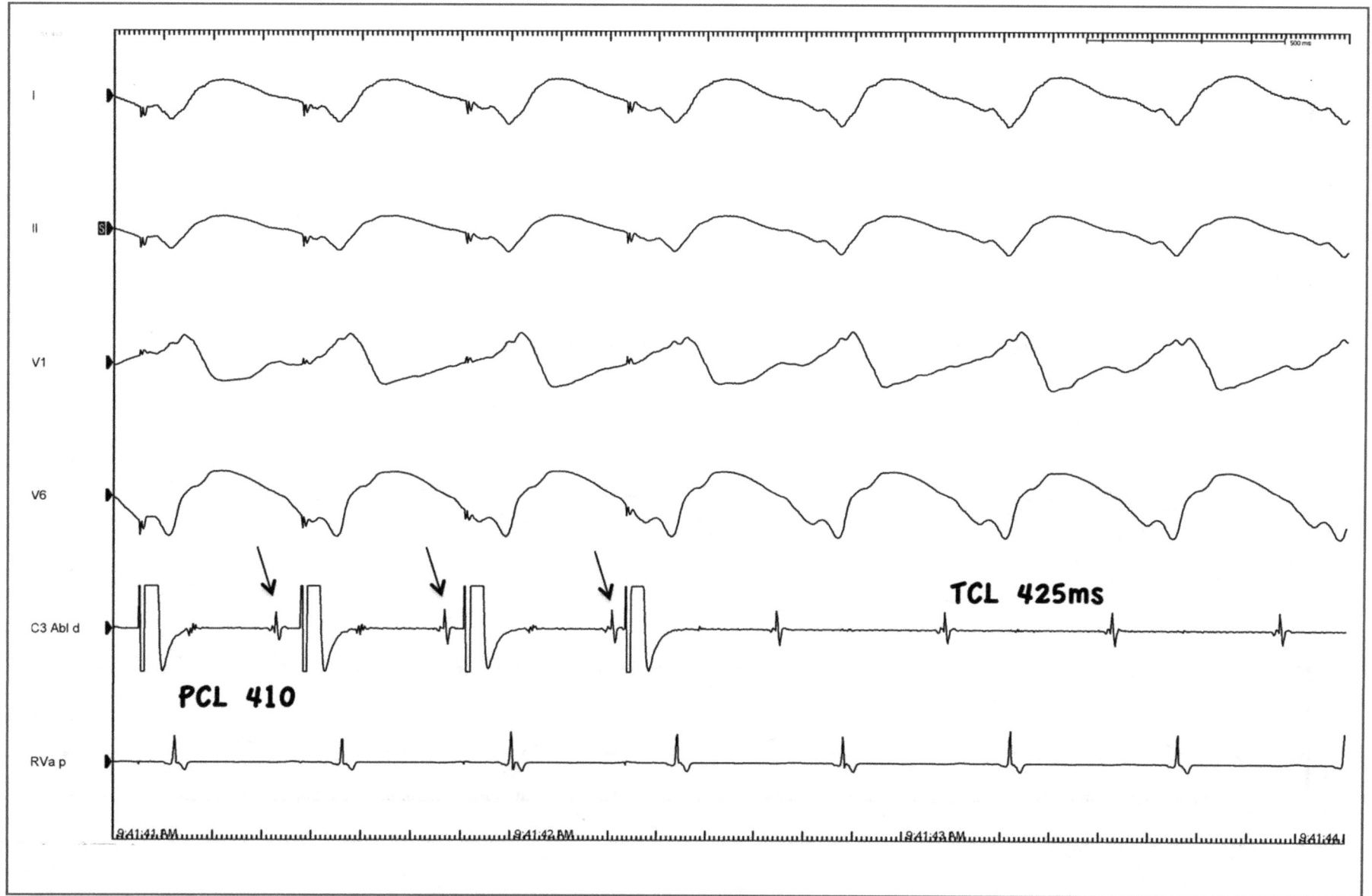

Figure 5.7

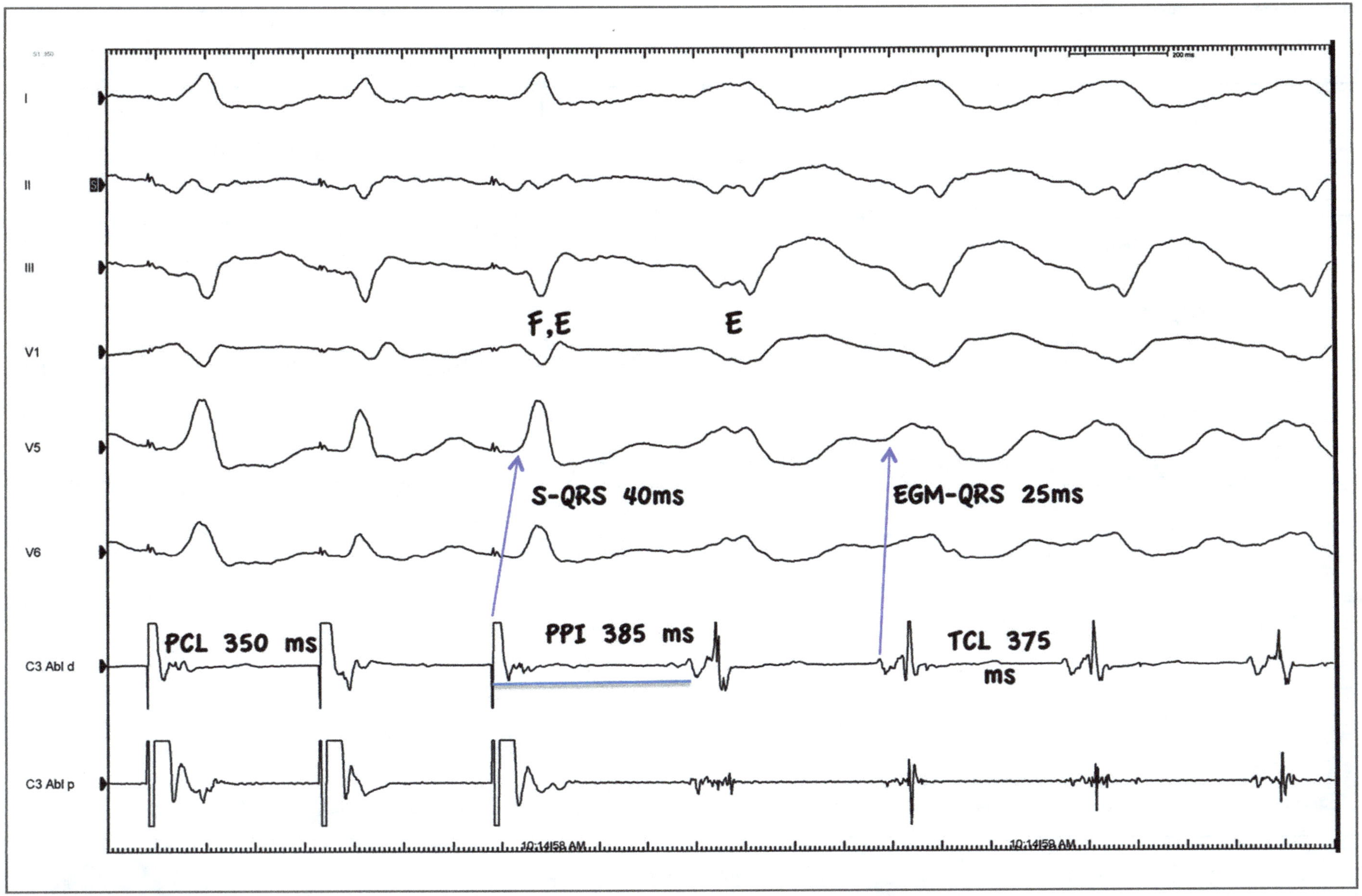

200 ms
I
II
III
F,E
E
V1
V5
S-QRS 40ms
EGM-QRS 25ms
V6
C3 Abl d
PCL 350 ms
PPI 385 ms
TCL 375 ms
C3 Abl p
10:14:58 AM
10:14:59 AM

**Figure 5.8**

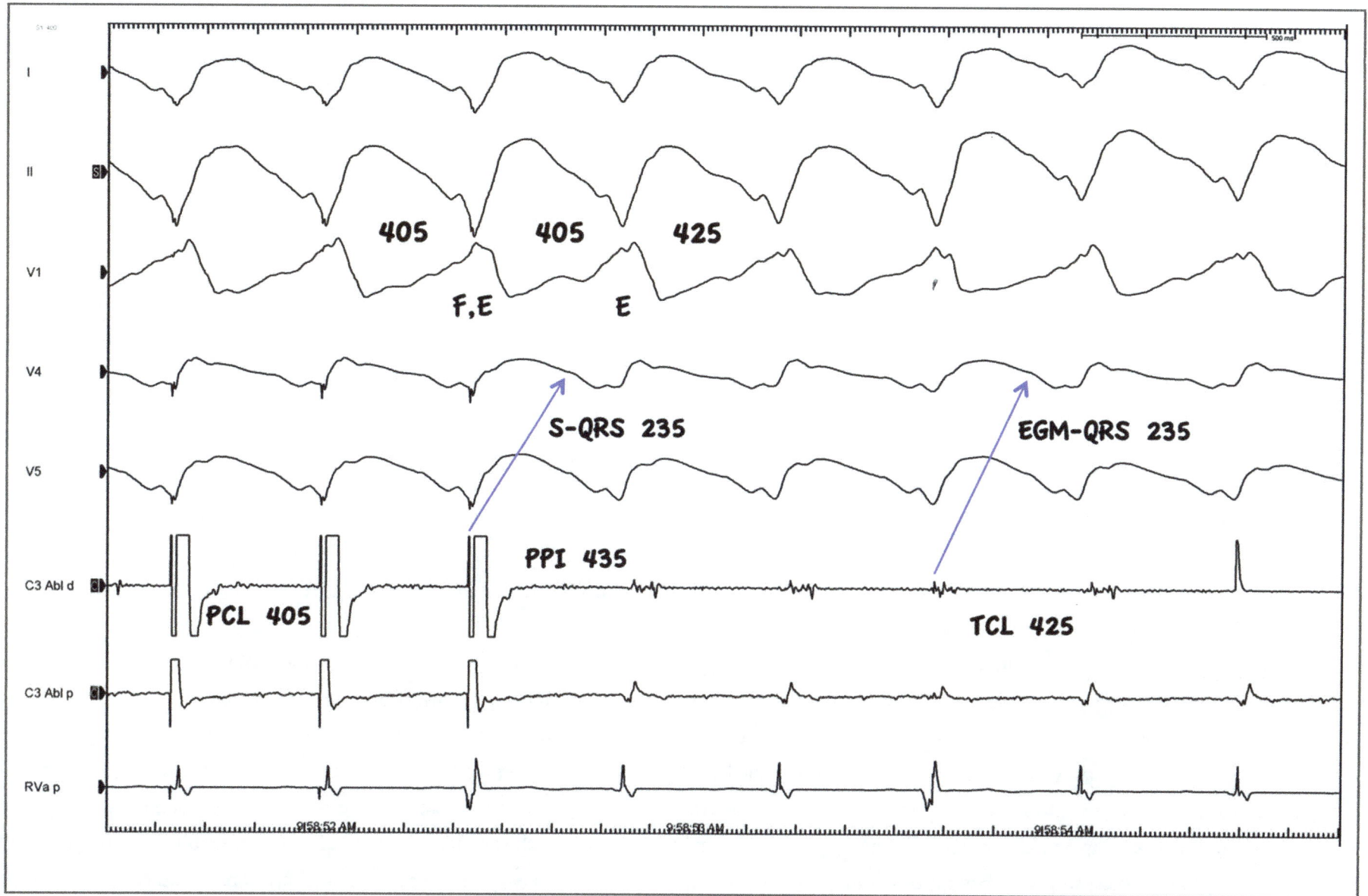

A complex EGM is recorded at the ablation site during VT in **Figure 5.9A**, the components of which are arbitrarily designated as 1, 2, and 3. Which component should be considered the one being captured (i.e., the "near-field" EGM) and which should be considered "far-field"? Any component occurring slightly prior to the stimulus artifact during ventricular pacing CANNOT be the one stimulated since it is unaffected by the pacing stimulus. This eliminates EGMs 1 and 2. When EGM 3 is used for measurement of the PPI (asterisk), the PPI is identical to the CL. In addition, the S-QRS = EGM-QRS, which is ~50% of the TCL, suggesting that the pacing site is near the central part of the critical isthmus (#2 on the schema).

**Figure 5.9B** is identical to Figure 5.9A, but is reproduced to illustrate the "$N+1$" concept. Although not the case here, in some instances, the pacing stimulus artifact may distort the EGM of the cycle paced and part of the postpacing EGM. In such an eventuality, one can use the *next* cycle for measurement, that is, the stimulus to the second QRS (S-QRS $(_{N+1})$) and the local EGM to the subsequent QRS (EGM-QRS $(_{N+2})$). Subtracting 1 CL from these will obviously provide the same measurements as in Figure 5.9A.

**Figure 5.9A**

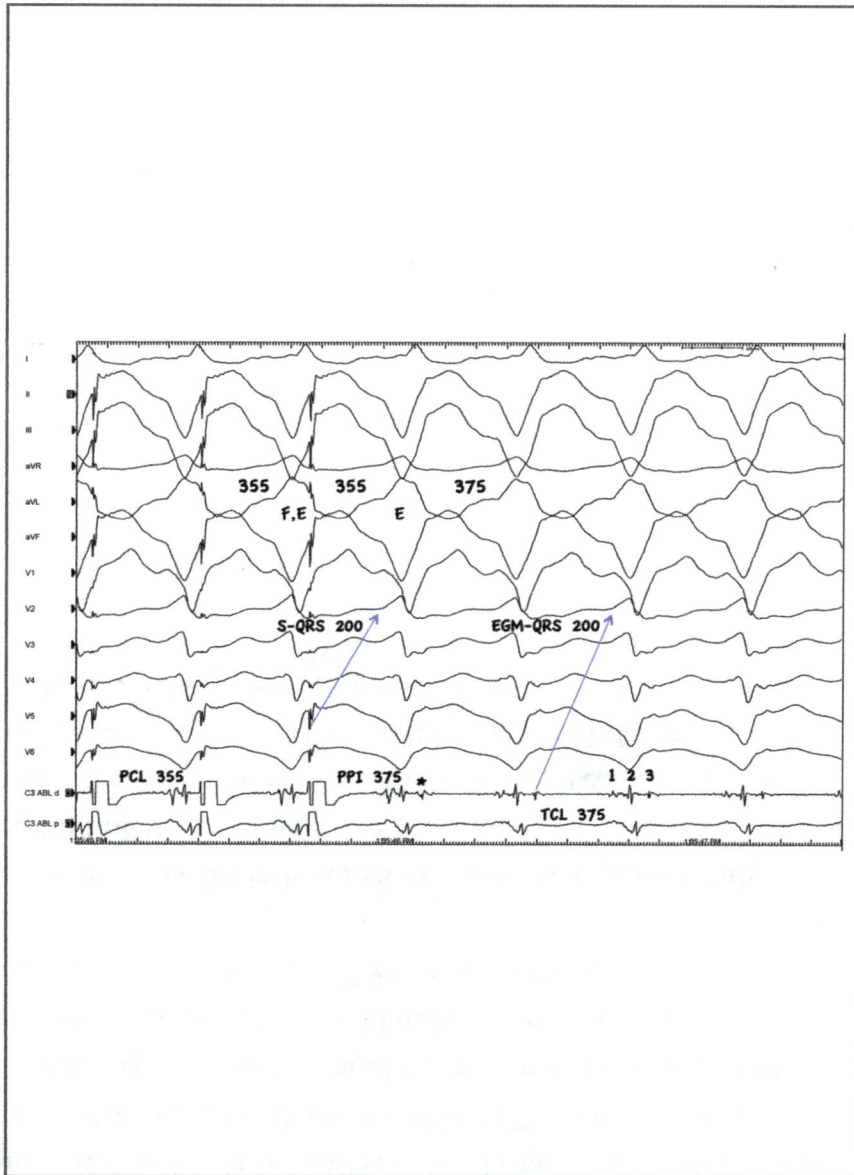

**Figure 5.9B**

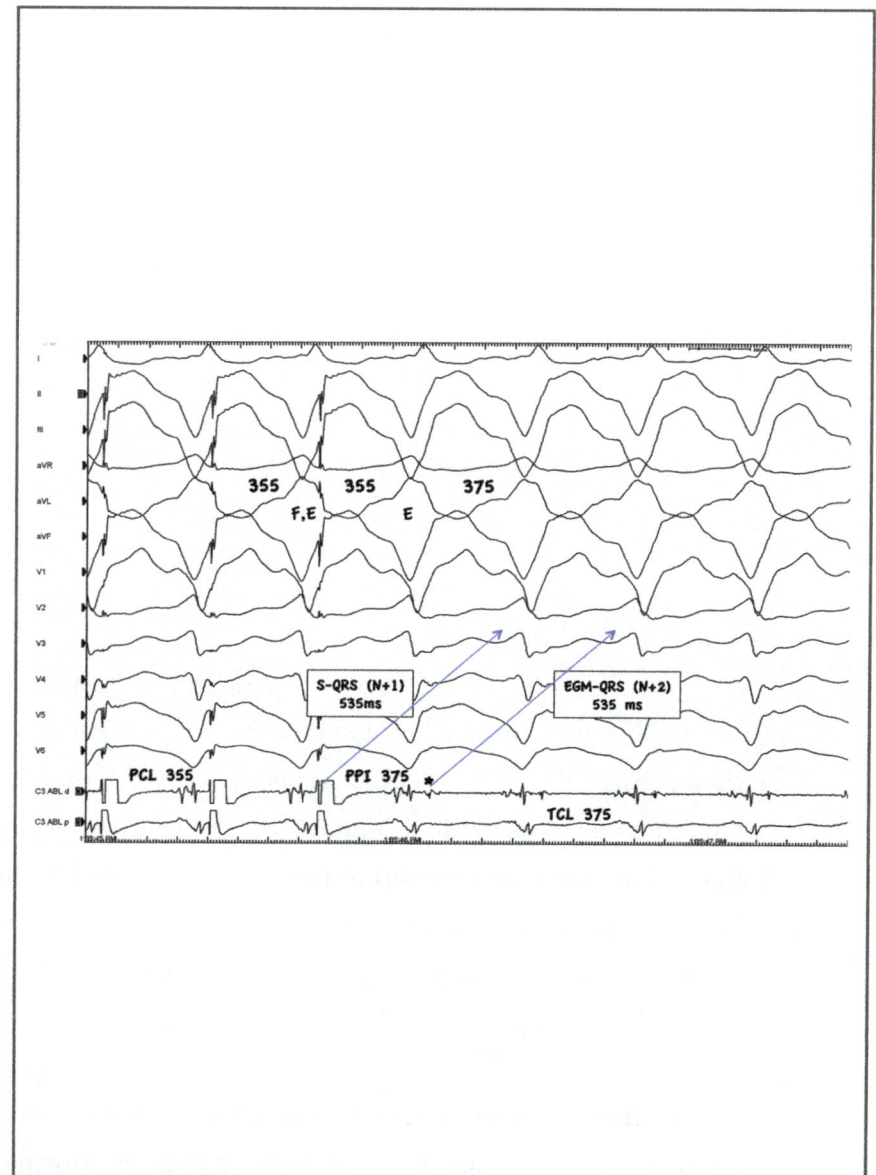

Pacing at the site in **Figure 5.10** entrains the tachycardia with concealed fusion and the PPI is "in" (PPI-TCL = 5 ms), supporting that the pacing site is in the central common pathway of the circuit. A very short S-QRS (21% of TCL) indicates that the catheter is at an isthmus site close to the exit of the scar (#1 in the schema).

Entrainment at the site in **Figure 5.11** entrains tachycardia with concealed fusion, again suggesting that we are in a protected diastolic corridor within the scar. A PPI-TCL of 0 confirms that the pacing site is in the circuit. The S-QRS and the EGM-QRS are both relatively long, taking up much (60%) of the TCL. This supports the stimulation site being within the obligatory circuit, likely more proximal in the circuit near an entrance site far from the exit (#3 in the schema).

Pacing at this site in **Figure 5.12** entrains with concealed fusion. However, a PPI-TCL of 70 ms and a S-QRS > EGM-QRS suggests that the pacing site is well out of the obligatory circuit, possibly in a blind loop adjacent to the circuit (#5 in schema). Note that the far-field diastolic potential (asterisk) remains unperturbed despite acceleration of the tachycardia during entrainment, suggesting that it is not critical to the circuit.

**Figure 5.10**

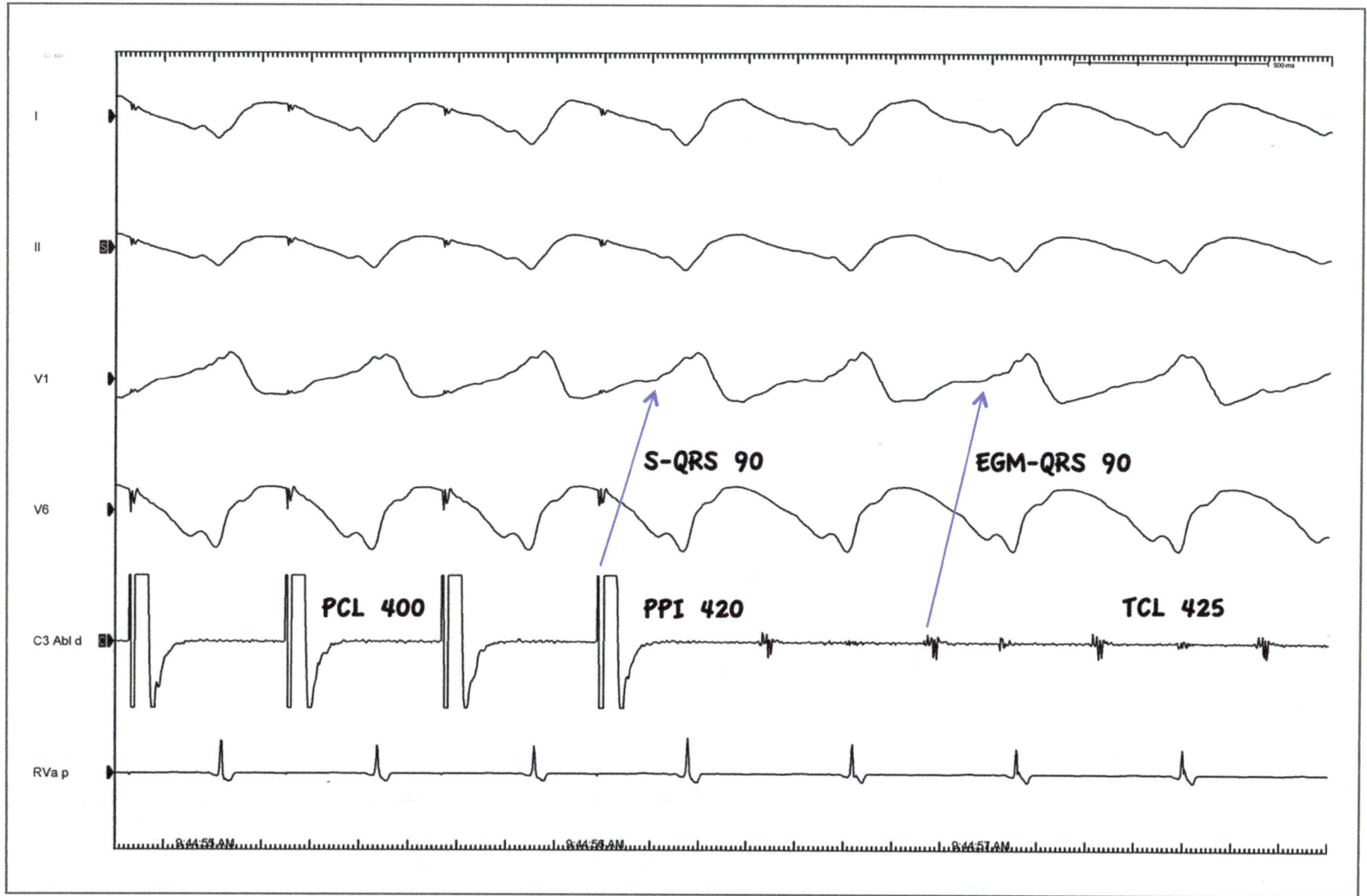

Figure 5.11

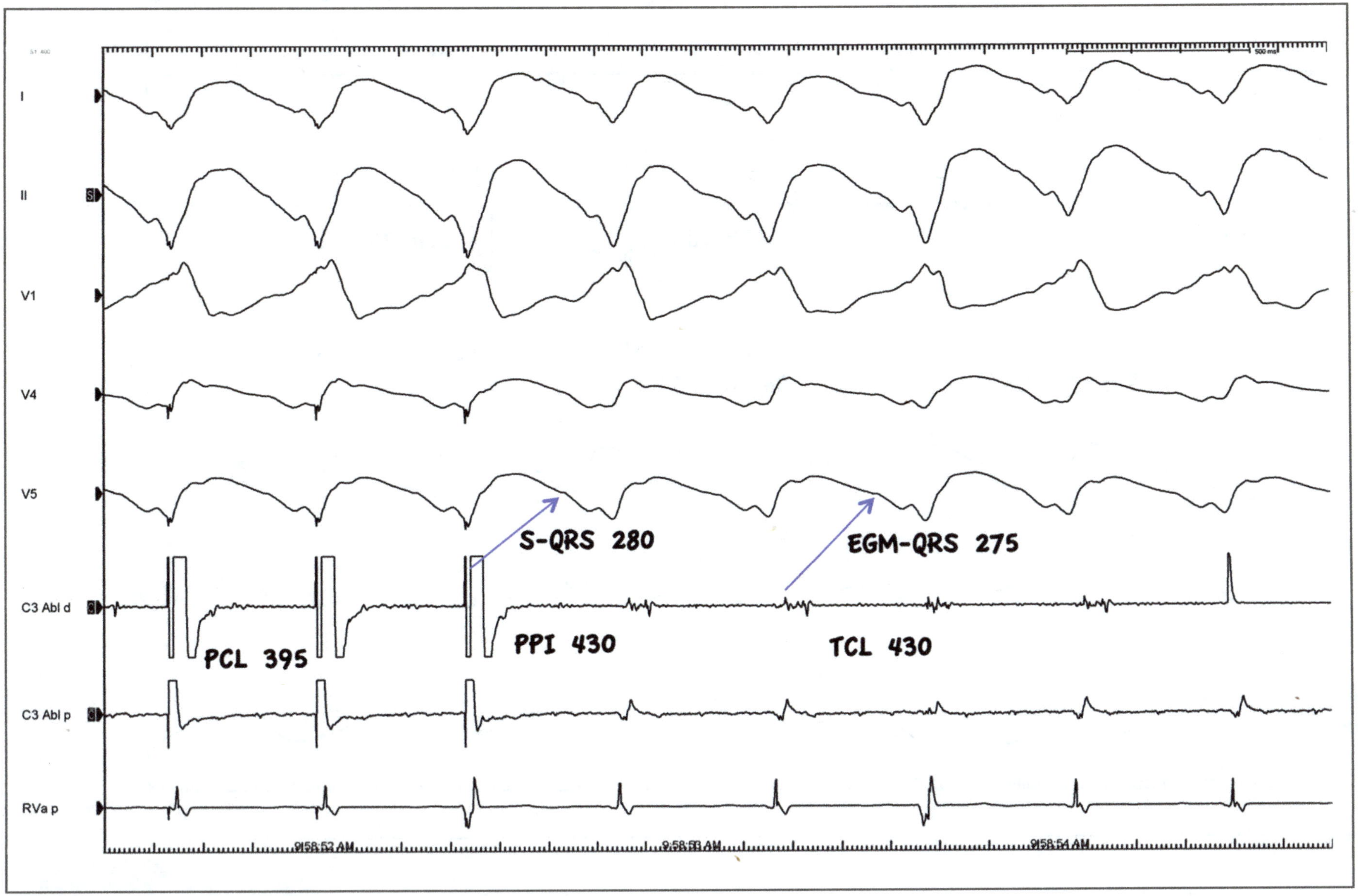
I
II
V1
V4
V5
C3 Abl d
C3 Abl p
RVa p
S-QRS 280
EGM-QRS 275
PCL 395
PPI 430
TCL 430
500 ms
9:58:52 AM
9:58:53 AM
9:58:54 AM

**Figure 5.12**

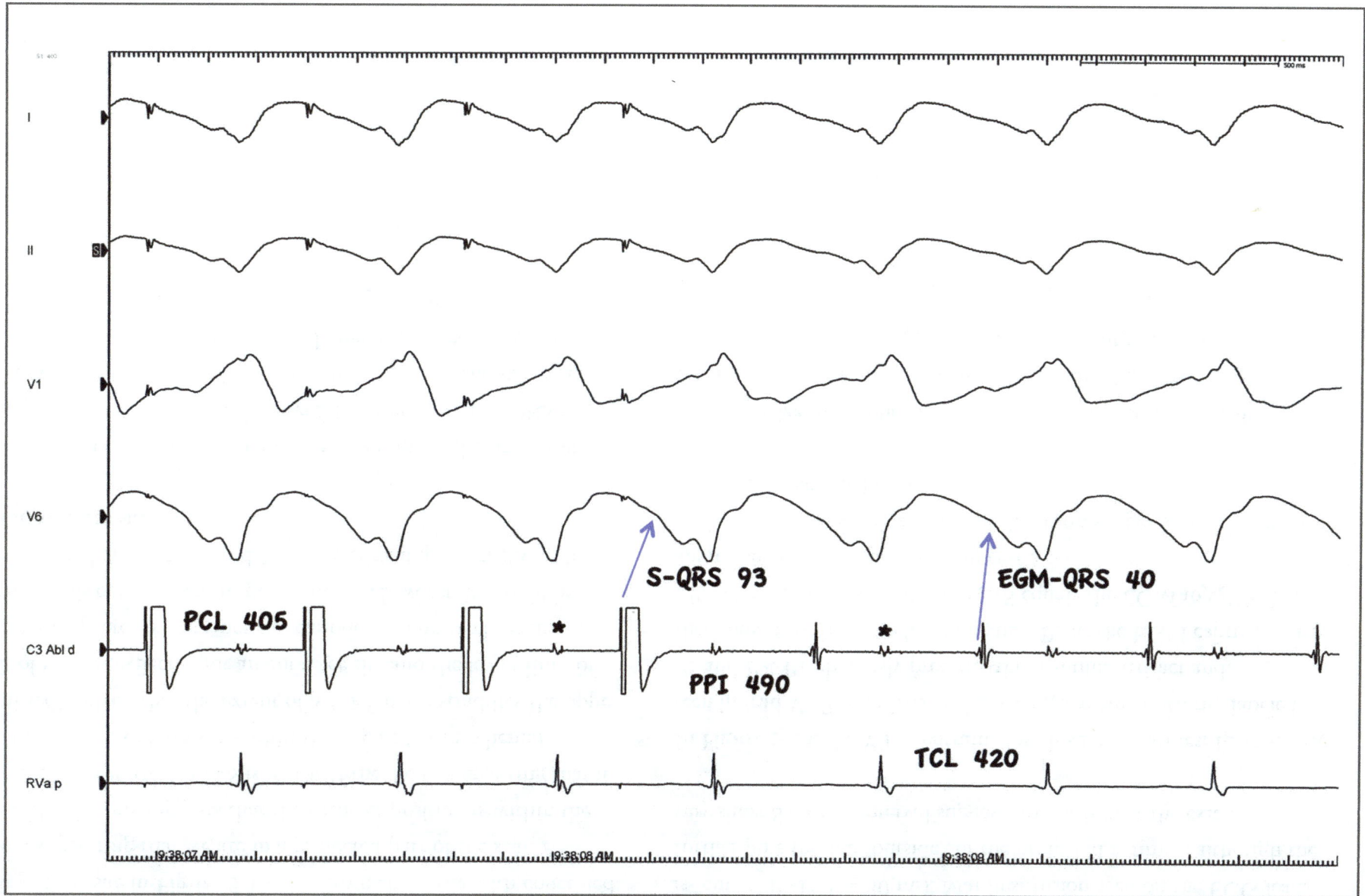

Pacing at the site in **Figure 5.13** entrains tachycardia with concealed fusion, suggesting that we are in a protected part of the scar. A PPI-TCL of 10 ms suggests that the catheter position is within the circuit. However, the S-QRS is ~70% of the TCL, suggesting that it may be at an inner loop site within the circuit (#4 in schema). Admittedly, given that the extent of S-QRS delay straddles the upper limit of what is expected for an entrance site and the lower limit of an inner loop site, an entrance site cannot be definitively excluded based on this entrainment response alone. However, it should be emphasized that site #4 would NOT be critical to the tachycardia circuit, whereas site #3 would be.

**Figure 5.14** illustrates fractionated electrograms at the pacing site during tachycardia. The first (arrow) follows onset of the QRS is probably the "near-field" EGM that is being paced. The subsequent ones (asterisks) are not consistent and obviously dissociated and unrelated to the tachycardia circuit. Pacing this site yields a PPI that is "out" (PPI-TCL = 80 ms). Manifest fusion seen on the ECG leads further puts the site "outside" of the protected isthmus, although the very short ST-QRS interval suggests that it is near the exit site.

In **Figure 5.15**, the VT is entrained with subtle manifest fusion, best seen in lead $V_4$. The EGMs at the pacing site are arbitrarily labeled P1 and P2. P1 obviously precedes the stimulus artifact and, therefore, must be far field. If one uses P2 as the EGM captured, the PPI is "in" and the stimulus to QRS equals the EGM to QRS (both approximately 0). This would suggest a site at or very close to the exit of the scar (#1 in schema). The subtle fusion suggests that the site is most likely just outside an exit site.

To emphasize the applicability of classical entrainment to this tracing, it is noted for clarity that QRS 1 and 2 are the last paced and entrained whereas QRS 3 is "entrained but not fused."

**Figure 5.13**

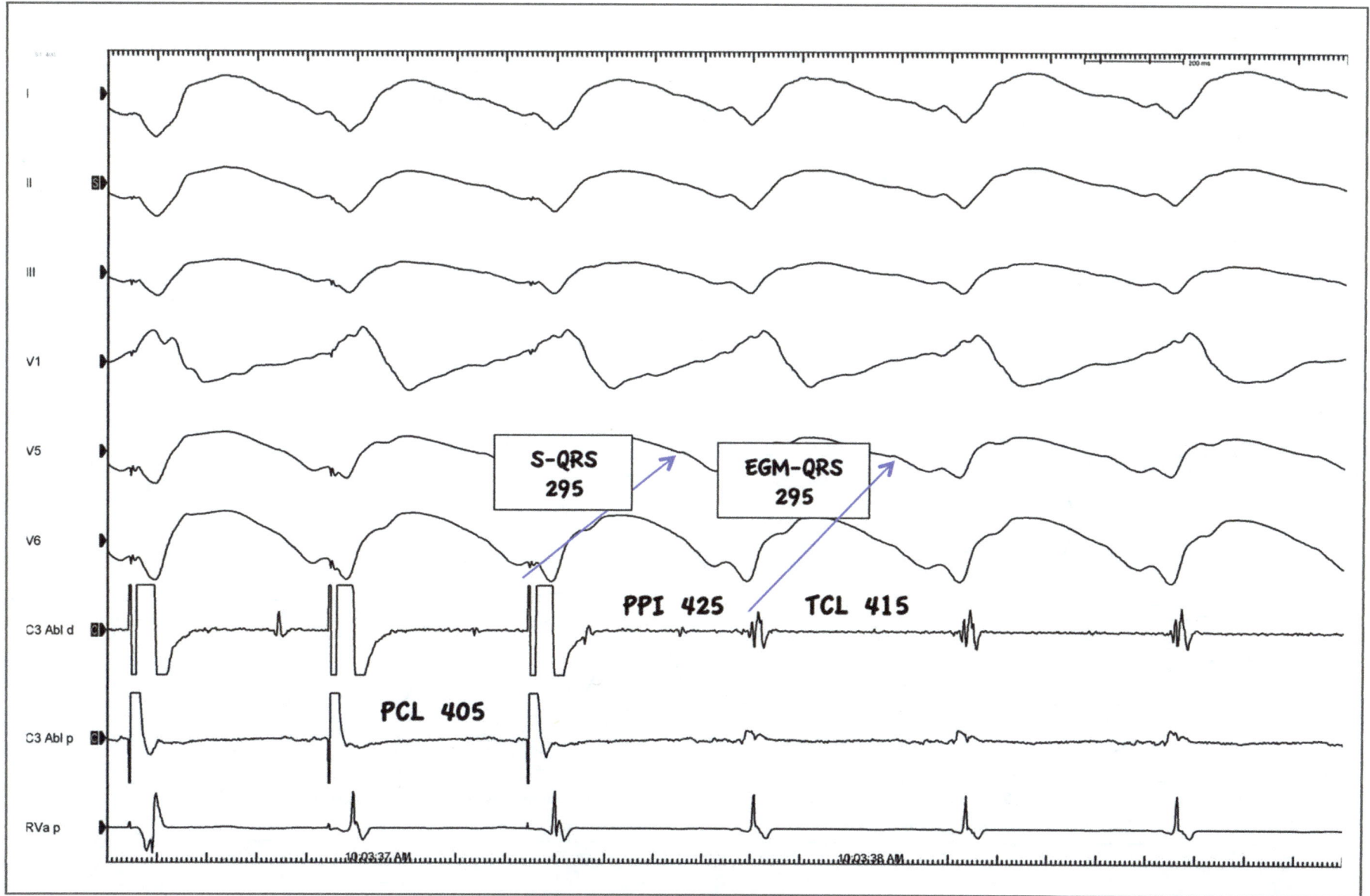

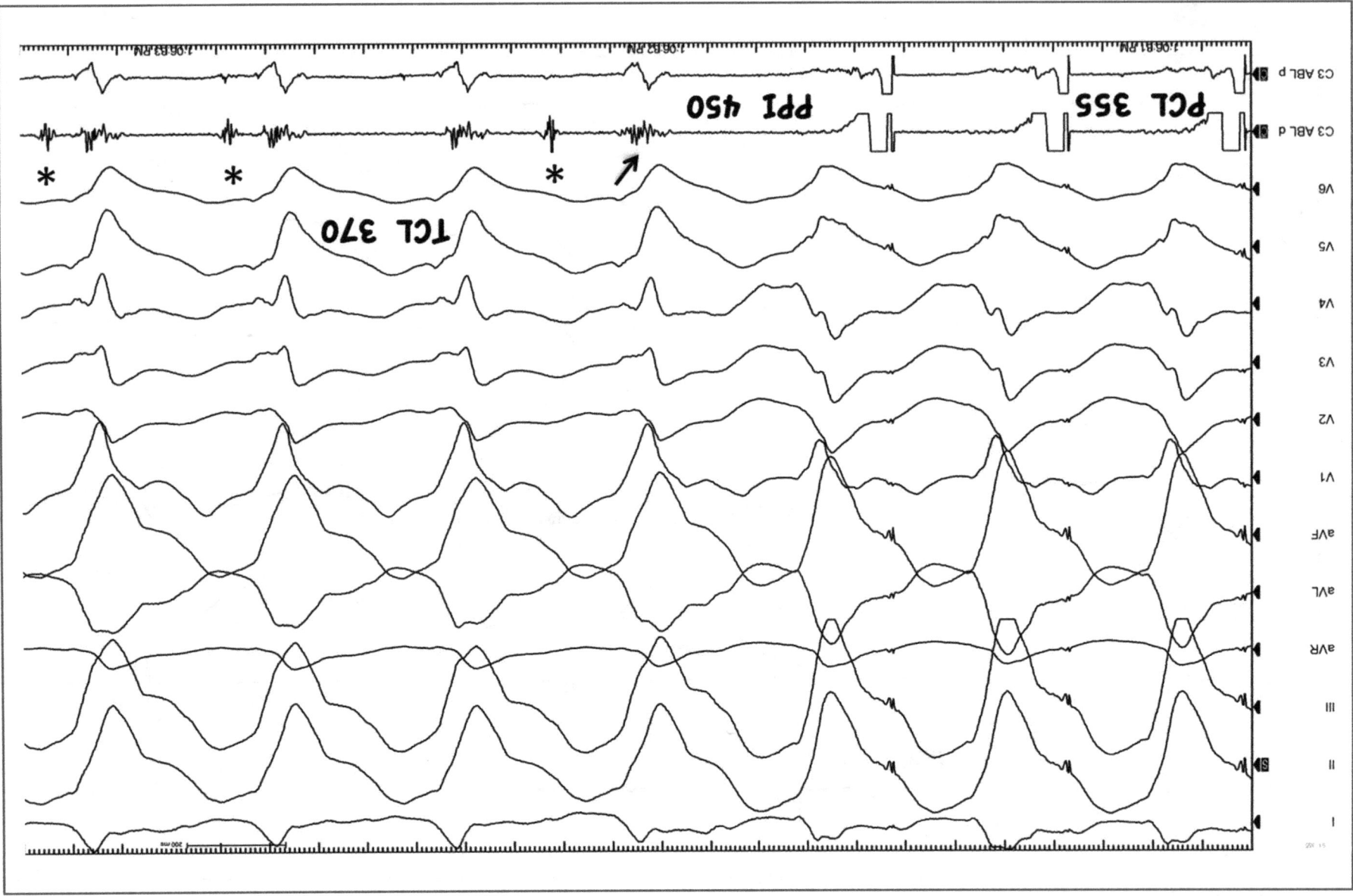

Figure 5.14

**Figure 5.15**

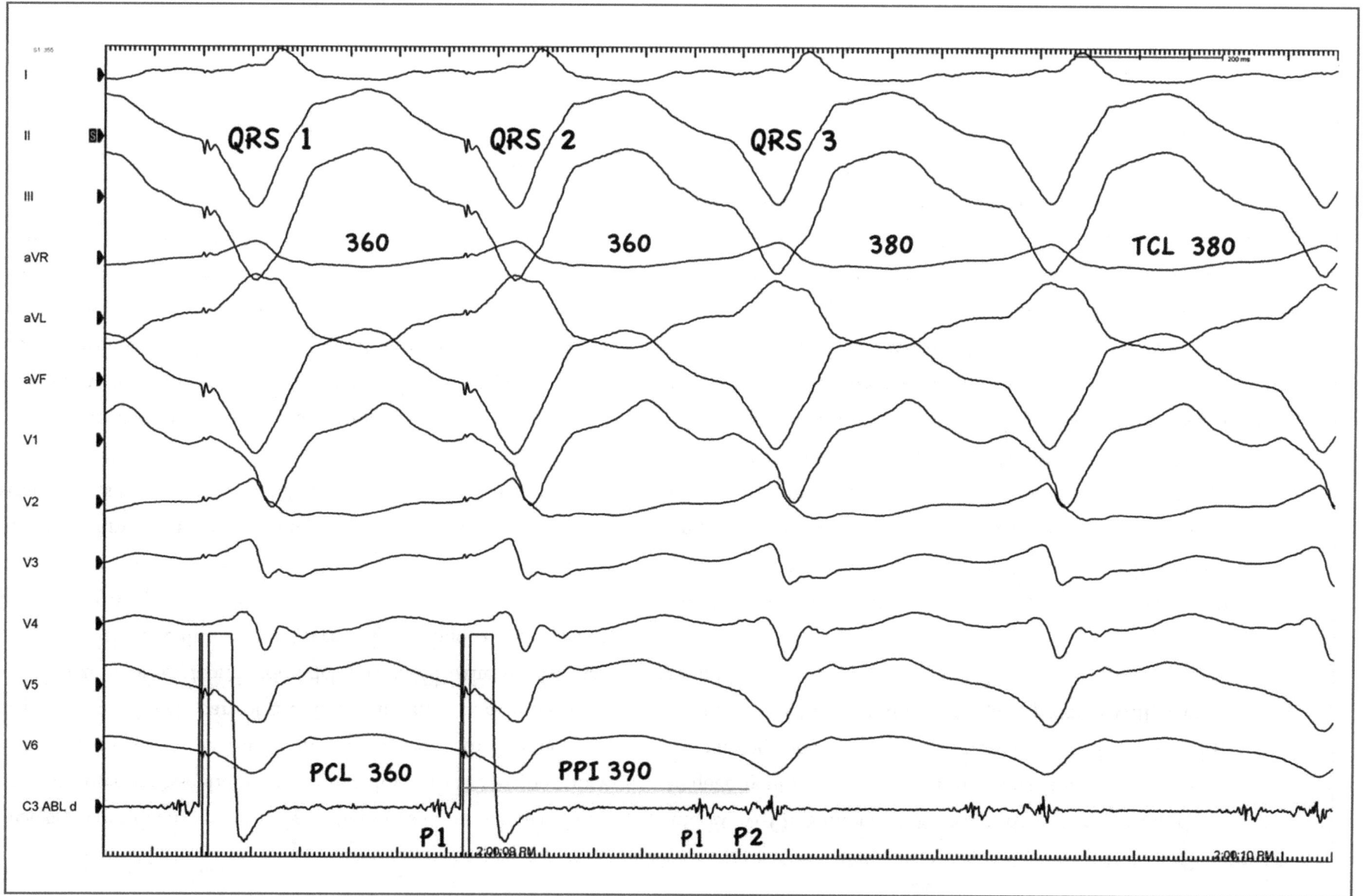

# Atrial Tachycardia

The relevant considerations for AT and flutter as far as our mandate is in this chapter are essentially the same as for VT. The major utility of entrainment is with "macrorentrant" AT to define the circuit and, in particular, a "critical" part of the circuit that may be a "protected isthmus" or zone of scar/conduction delay as in VT. Alternately, the critical part of the circuit may not necessarily be abnormal tissue but a relatively narrow part of the circuit that is more amenable to ablation such as the tricuspid-caval isthmus in typical atrial flutter or the "mitral isthmus" in perimitral reentry. Much of the atrial substrate comes from pathology related to atrial fibrillation itself and its subsequent ablation and to congenital or other cardiac disease further complicated by operative management. Semantics often confuse but we will consider atrial "flutter" to be a subtype of AT, historically differentiated by rate (i.e., atrial flutter generally being any AT in the range of ≥ 240 bpm). Of course, we appreciate that this distinction is largely arbitrary since atrial rate alone does not distinguish tachycardia mechanism.

The QRS morphology during VT as noted above is very useful in assessment of manifest vs. concealed fusion and determining S-QRS and EGM-QRS intervals. P-wave morphology during the ATs is frequently of low amplitude and less useful. Although P-wave morphology generally cannot be used as in VT, atrial EGMs can be used to assess fusion and reset to good profit as will be evident from examples.

Focal AT related to automaticity or triggered activity cannot be "entrained." Microreentrant AT can be theoretically entrained but progressive fusion may be hard to demonstrate. Regardless of mechanism, focal AT can be overdriven with atrial pacing. Theoretically, *the return cycle or PPI at the pacing site should have a direct relationship to the distance of the pacing catheter from the focus, given a constant-pacing CL.* This principle has been demonstrated (see Mohamed et al, Suggested Readings) to be useful in localization of a focus. The PPI represents the sum of the conduction time from the pacing site to the perifocal tissue, the time required to penetrate the perifocal tissue to reset the focus, the CL of the focus, and the conduction time back to the pacing site.

The perifocal conduction time for foci may be variable and need not be 0 even at an optimal site since conduction time through the perifocal tissue may be variable. Consider the sinus node as a "model" focus where the recovery time after atrial pacing requires a time factor to reach the sinus node, traverse the perisinus node tissue, reset the sinus node, and return to origin. In the case of the sinus node, the "perifocal" conduction time will often be in the range of ≥ 100 ms. Nonetheless, the *shortest PPI* recorded should approximate the focus site if adequate points are sampled. The example in **Figure 5.16** illustrates a high right atrial focus (arrow on central point in red on the electroanatomic map) that corresponds to the shortest PPI recorded after atrial mapping (star symbol AT). This was the successful ablation site.

Typical isthmus-dependent atrial flutter provides an excellent model of a macroreentrant circuit and a useful template to aid in interpreting unknown atrial circuits.

**Figure 5.17** is such a case with the pacing site (ABLd) located at the tricuspid-caval isthmus. A multielectrode (Halo®) is positioned as usual seated within the RA and around the tricuspid valve with the distal poles near the isthmus (electrodes LRA 1–2 to HRA 19–20).

The first observation is that the right atrial intracardiac EGMs suggest a circulating wave moving counterclockwise around the tricuspid valve. The right EGMs span virtually the whole cardiac cycle (206 ms), a finding very suggestive of macroreentry in the RA.

The most important point, from a practical point of view, is that the tachycardia has been properly entrained and the PPI = TCL. This confirms that the isthmus (the usual ablation target) is part of the circuit. Additionally, the P wave is relatively well visualized here and the paced P wave is virtually identical to the tachycardia P waves recorded (as verified by the full ECG (not shown)). Furthermore, the Stim-P and EGM-P are equal, further supporting that the site is in the circuit. The intracardiac EGMs have very similar activation patterns, which also support the site being in the circuit. A minor degree of fusion is noted in the intracardiac EGMs, best appreciated at the CS EGMs (asterisk).

**Figure 5.16**

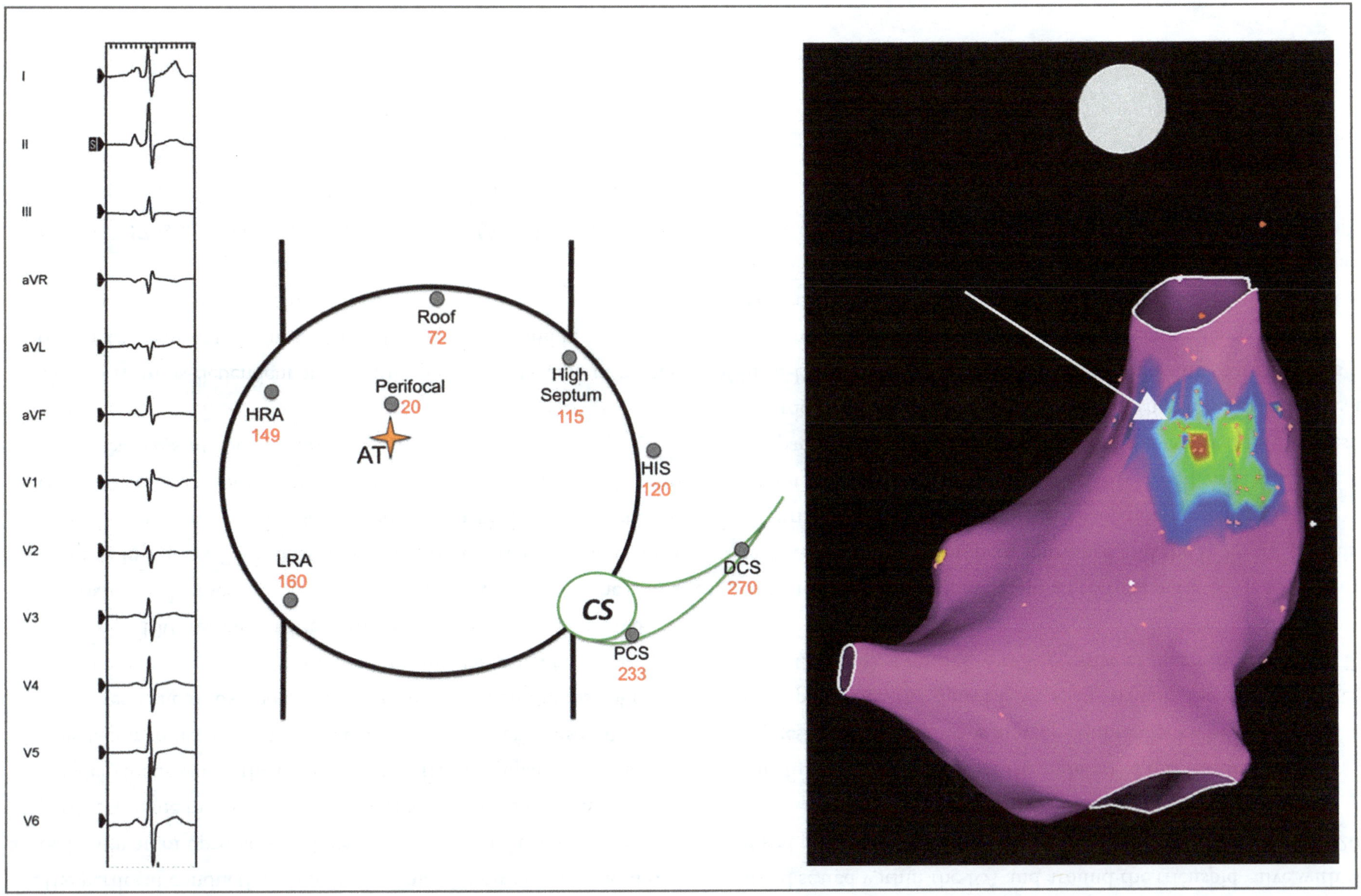
I
II
III
aVR
aVL
aVF
V1
V2
V3
V4
V5
V6
Roof
72
Perifocal
20
High
Septum
115
HRA
149
AT
HIS
120
LRA
160
DCS
270
CS
PCS
233

**Figure 5.17**

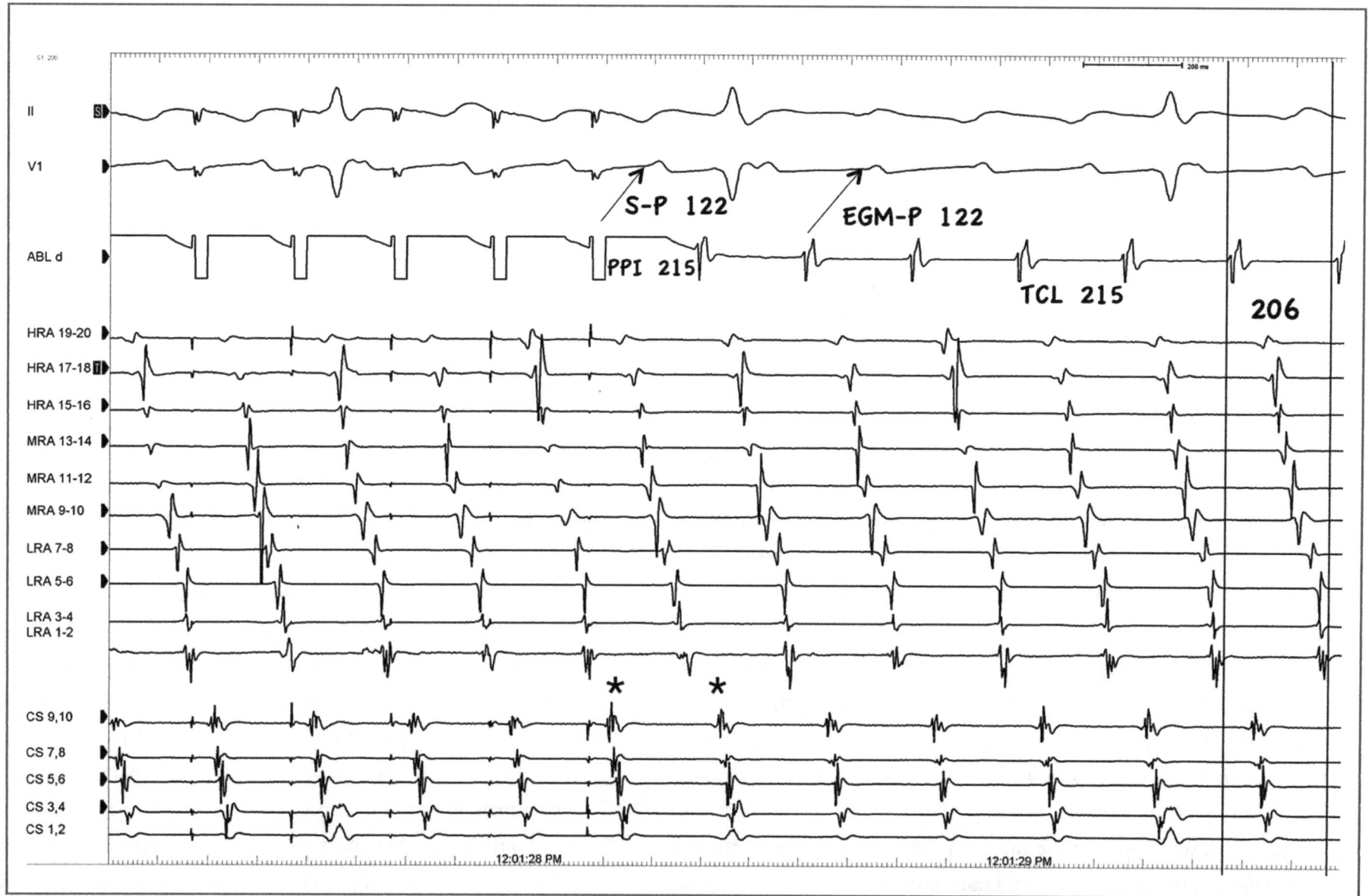

**Figure 5.18** is recorded from another individual with typical atrial flutter and the electrodes are laid out similarly. It is again observed that most of the cardiac cycle is recorded in the right atrium as would be expected with RA macroreentry. The proximal CS electrode is slightly within the orifice of the CS. Pacing at a distance from the macroreentrant circuit maximizes the opportunity for observing fusion that is readily apparent from the intracardiac EGMs. The LA, represented by CS 9–10, measures a PPI of 360 ms, clearly out of the circuit (as is expected).

The strategy for typical flutter is essentially the same for any stable atrial macroreentrant tachycardia. **Figures 5.19** and **5.20** were from a patient with remote operative therapy of a right atrioventricular pathway. In Figure 5.19, the pacing catheter is near and medial to the tricuspid-caval isthmus. The right atrial EGMs cover most of the cardiac cycle, suggesting right atrial macroreentry. However, the PPI from the site is very long, a finding that would not be compatible with tricuspid-caval isthmus-dependent atrial flutter. In addition, the EGMs during entrainment clearly show manifest fusion. In contrast, the site in Figure 5.20 is low lateral RA. The PPI at this site equaled the TCL, and the atrial activation sequence during entrainment was very similar to tachycardia, approximating "entrainment with concealed fusion." This was a successful ablation site and was presumably related to scar from atrial surgery many years previously.

**Figure 5.18**

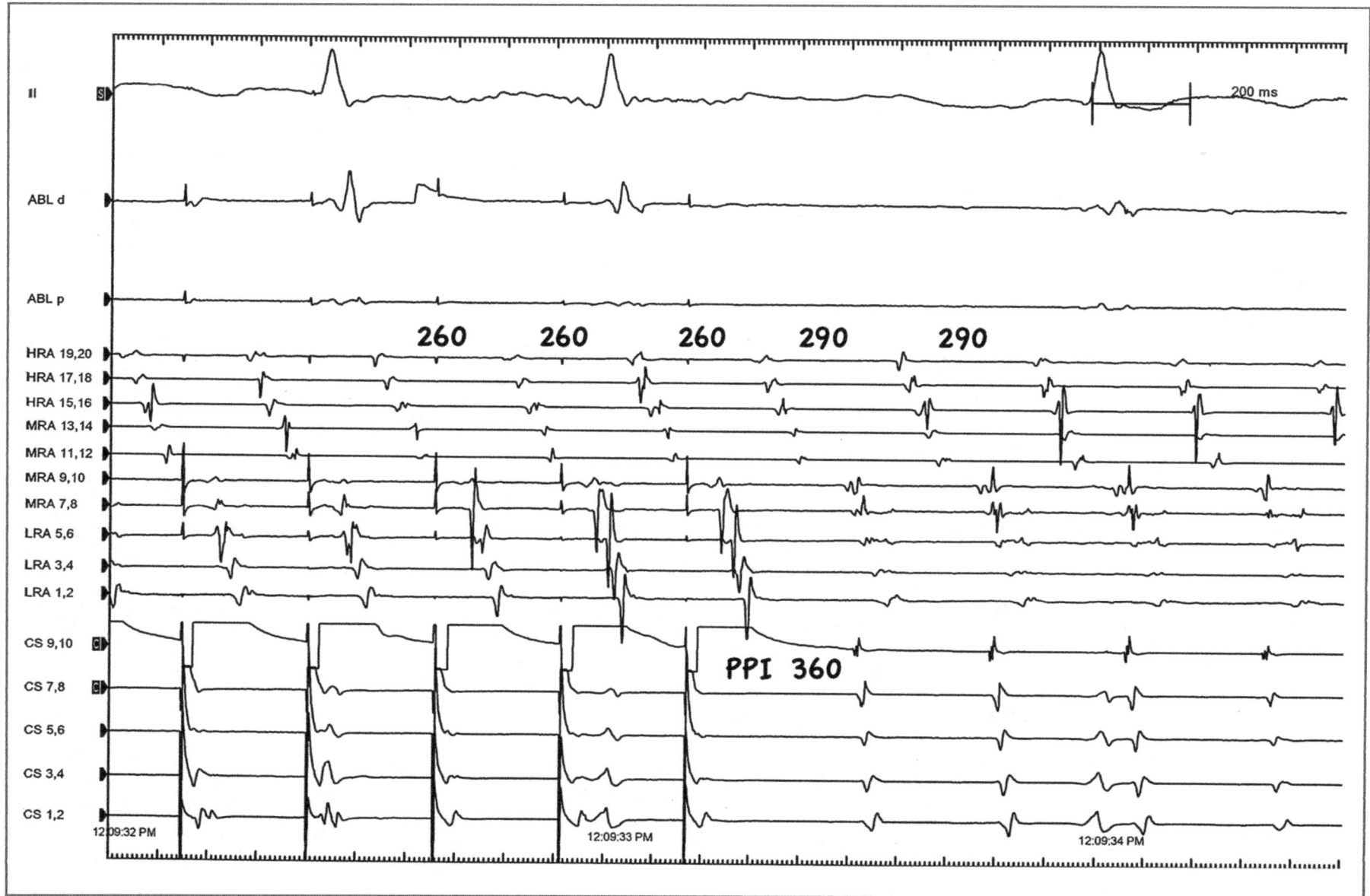

Figure 5.19

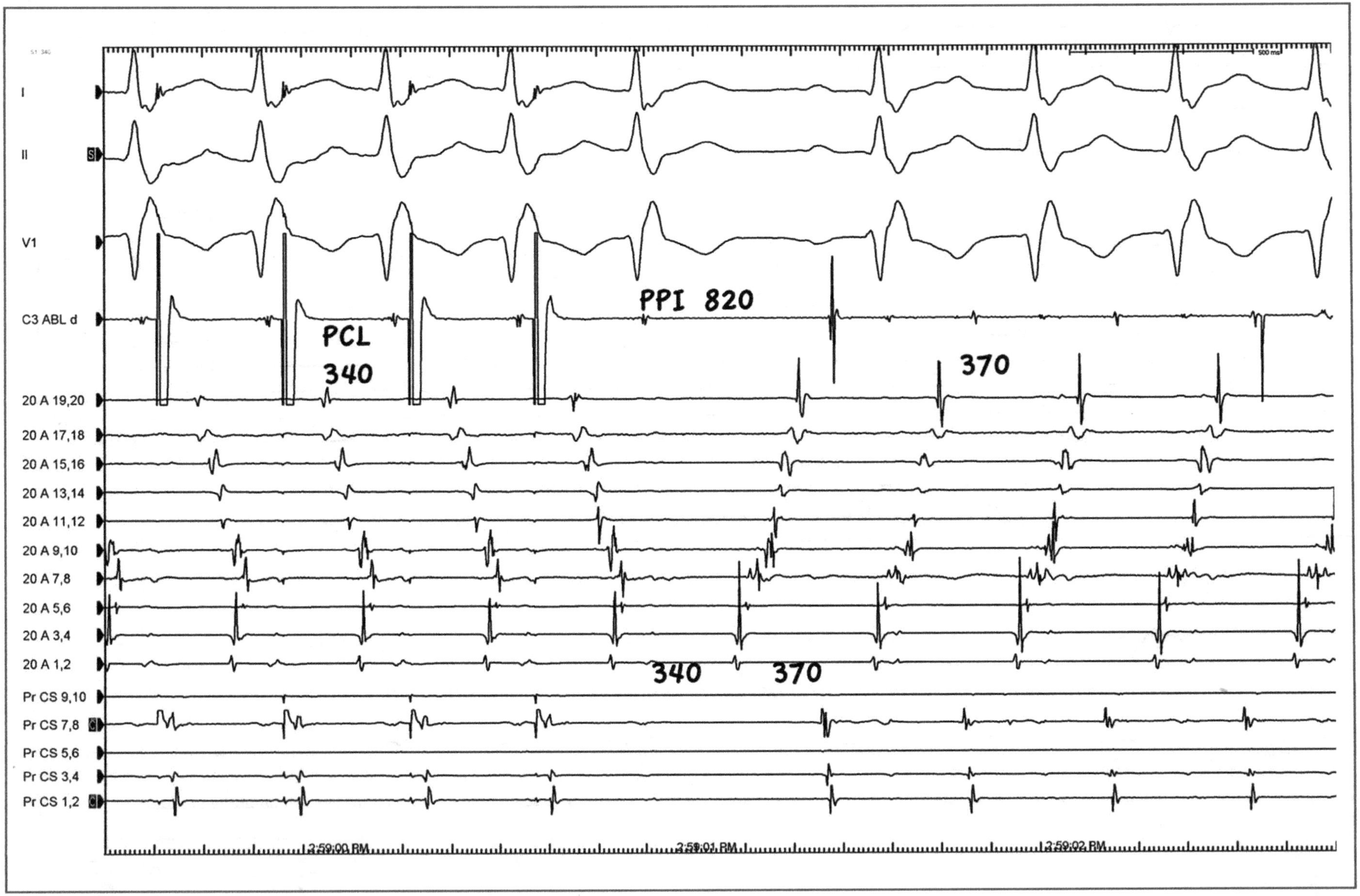
500 ms
I
II
V1
C3 ABL d
PPI 820
PCL
340
370
20 A 19,20
20 A 17,18
20 A 15,16
20 A 13,14
20 A 11,12
20 A 9,10
20 A 7,8
20 A 5,6
20 A 3,4
20 A 1,2
340
370
Pr CS 9,10
Pr CS 7,8
Pr CS 5,6
Pr CS 3,4
Pr CS 1,2
2:59:00 PM
2:59:01 PM
2:59:02 PM

**Figure 5.20**

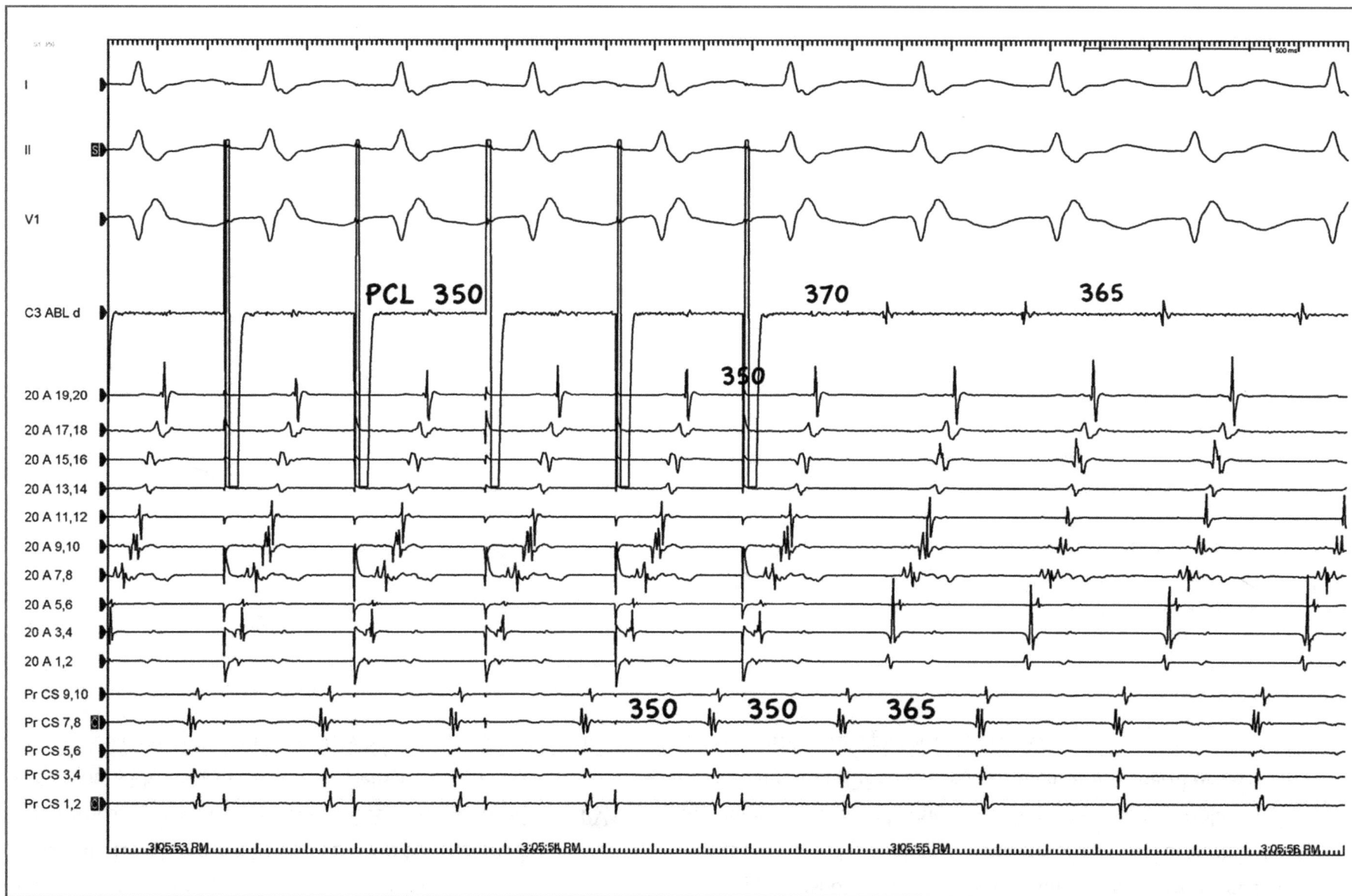

The patient relating to **Figure 5.21** had AT after pulmonary vein (PV) ablation. The electrodes labeled "Ls" are from a circular mapping catheter in the right inferior PV. A close PPI inside the PV in conjunction with late PPI outside the vein allowed a clear diagnosis of AT coming from that vein, which required reisolation.

PPI mapping is, of course, very useful for stable AT from the LA as well as the right. For example, a pattern of PPI "in" at both the anterior and posterior LA near the ring is most compatible with perimitral flutter. Of course, entrainment is less useful or not even feasible with more rapid or irregular ATs.

**Figure 5.21**

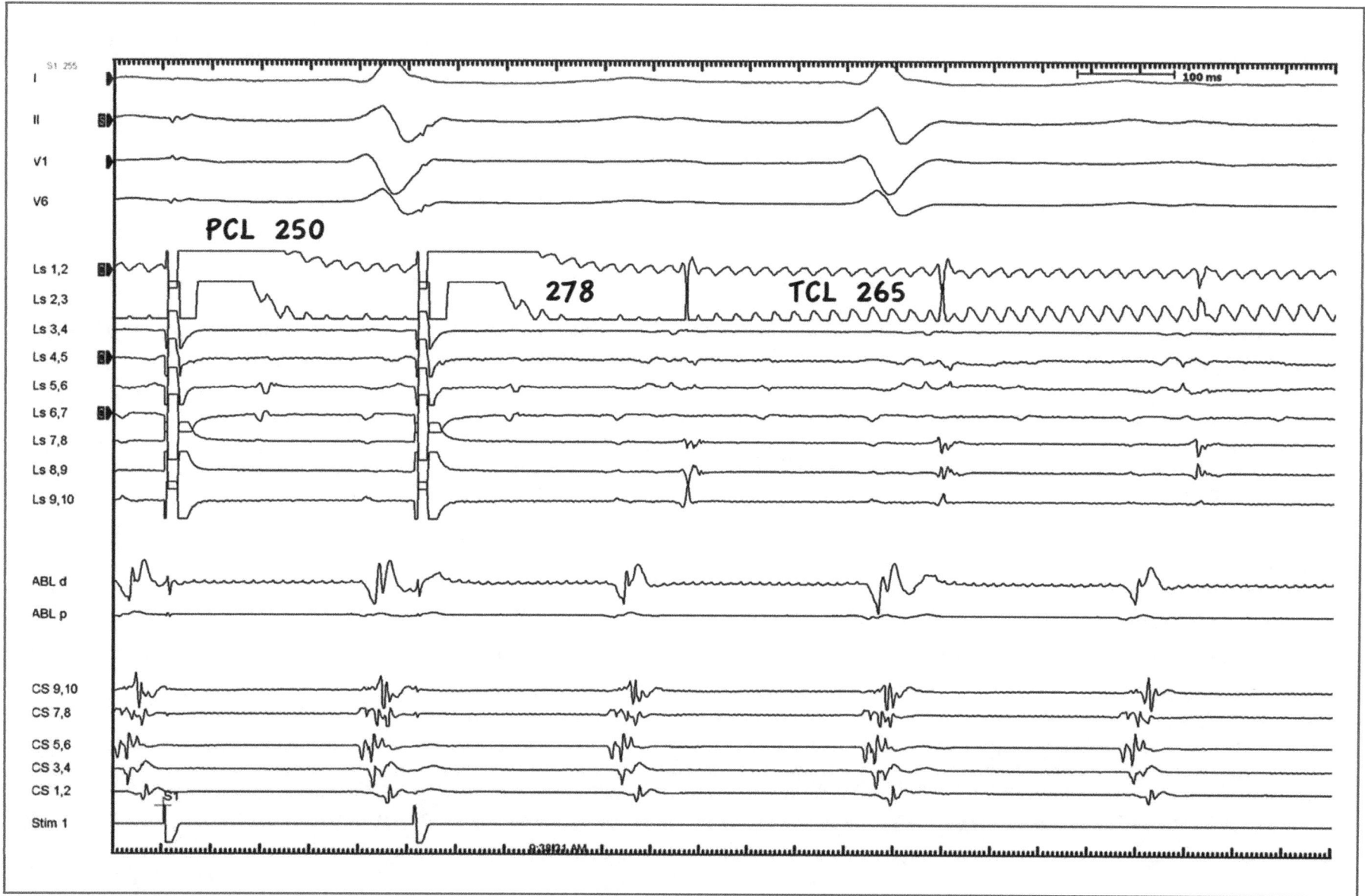

# Suggested Readings

1. Stevenson WG, Khan H, Sager P, et al. Identification of reentry circuit sites during catheter mapping and radiofrequency ablation of ventricular tachycardia late after myocardial infarction. *Circulation*. 1993;88:1647–1670.

2. Soegima K, Stevenson WG, Maisel WH, et al. The N + 1 difference: A new measure for entrainment mapping. *J Am Coll Cardiol*. 2001;37:1386–1394.

3. Stevenson WG, Friedman PL, Sager PT, et al. Exploring post infarction reentrant ventricular tachycardia with entrainment mapping. *J Am Coll Cardiol*. 1997;29:1180–1189.

4. Delacretaz E, Stevenson WG. Catheter ablation of ventricular tachycardia in patients with coronary heart disease. *PACE*. 2001;24(8):1261–1277.

5. Issa ZF, Miller JM, Zipes DP. Post-infarction sustained monomorphic ventricular tachycardia. In: *Clinical Arrhythmology and Electrophysiology: A Companion to Braunwald's Heart Disease*. 1st ed. Philadelphia, PA: Saunders Elsevier; 2009.

6. Jaïs P, Matsuo S, Knecht S, et al. A deductive mapping strategy for atrial tachycardia following atrial fibrillation ablation: Importance of localized reentry. *J Cardiovasc Electrophysiol*. 2009;20:480–491.

7. Weerasooriya R, Jaïs P, Wright M, et al. Catheter ablation of atrial tachycardia following atrial fibrillation ablation. *J Cardiovasc Electrophysiol*. 2009;20:833–838.

8. Miyazaki H, Stevenson WG, Stephenson K, Soejima K, Epstein LM. Entrainment mapping for rapid distinction of left and right atrial tachycardias. *Heart Rhythm*. 2006;3:516–523.

9. McElderry HT, McGiffin DC, Plumb VJ, et al. Proarrhythmic aspects of atrial fibrillation surgery: Mechanisms of postoperative macroreentrant tachycardias. *Circulation*. 2008;117:155–162.

10. Mohamed U, Skanes AC, Gula LJ, et al. A novel pacing maneuver to localize focal atrial tachycardia. *J Cardiovasc Electrophysiol*. 2007;18:1–6.

11. Bennett MT, Gula LJ, Klein GJ, Yee R, Krahn AD, Leong-Sit P, Skanes AC. An alternative method of assessing bidirectional block for atrial flutter. *J Cardiovasc Electrophysiol*. 2011;22(4):431–435.

# 6

# Nonstimulation Based Maneuvers

**P**acing maneuvers are routinely used and remain critical to induce and terminate tachycardia, and elucidate their mechanism in the electrophysiology (EP) laboratory. Nonetheless, pharmacological agents given intravenously during an EP study and other maneuvers remain useful. Drugs take advantage of the relative differential sensitivity of various cardiac tissues to an agent or class of drug. Indeed, there may be drugs very specific for given tissues that are as yet undeveloped. They are especially useful because they can also be employed outside the EP laboratory. Obtaining good quality continuous surface ECG recordings during drug administration with careful observation may be all that is required to establish a diagnosis in many cases.

# Isoproterenol

The major value of isoproterenol is to facilitate induction and/or maintenance of tachycardia for further assessment or ablation. Some arrhythmias are exquisitely sensitive to sympathetic tone so that lying supine, sedated, and immobile in the EP laboratory can impair induction and/or maintenance. Tachycardias due to abnormal or triggered automaticity, including idiopathic right ventricular (RV) and left ventricular (LV) tachycardia and some focal atrial tachycardias (AT), may require isoproterenol for induction, perhaps by enhancing phase 4 depolarization or inducing delayed afterdepolarizations.

Sympathetic tone may be critical to achieving the right balance of conduction velocity and refractoriness between the anterograde and retrograde limbs of the reentrant circuit in order to induce sustained tachycardia with atrioventricular nodal reentrant tachycardia (AVNRT) or atrioventricular reentrant tachycardia (AVRT). Usually, refractoriness of the anterograde limb (generally the atrioventricular (AV) node) is the main factor limiting maintenance of these tachycardias. Isoproterenol shortens the effective refractory period (ERP) of the limiting pathway sufficiently that it recovers in time to be activated by subsequent reentrant wave fronts. It may also facilitate the induction of many other arrhythmias including AT and ventricular tachycardia (VT) by various mechanisms.

# Adenosine

Adenosine is arguably the most widely used and useful intervention and emphasis in this section is appropriate. Any tachycardia dependent on AV nodal conduction is identified by its effect, prolonging or blocking AVN conduction. Focal tachycardias may be intrinsically adenosine sensitive.

The cardiac electrophysiologic effects seen with adenosine are initiated by binding to adenosine (A1) receptors. This activates an outward K-channel ($I_{KAch, Ado}$), which hyperpolarizes the transmembrane potential. In supraventricular tissue, this results in decreased sinus node automaticity, shortening of atrial refractoriness, and depressed AV node action potential generation that depresses AV nodal conduction (negative dromotropic effect). Hyperpolarization of the transmembrane potential can also restore or improve myocardial conduction in depressed tissue (tissue injured during radiofrequency ablation). Adenosine has minimal direct effects on ventricular myocardium and most accessory AV pathways (AP) but is a potent antiadrenergic agent. Blocking adrenergic activation of the transient inward and calcium currents ($I_{TI}$ and $I_{Ca}$) is responsible for adenosine's ability to terminate some idiopathic VT (specifically RVOT VT) but has no effect on others (idiopathic LV fascicular VT or scar-mediated VT).

Adenosine's characteristic effect of depressing or blocking AV nodal conduction makes it a useful diagnostic tool in a variety of clinical situations.

# Confirming Existence of a Tachycardia Substrate

In patients suspected of having an AP, ventricular preexcitation (delta waves) may be difficult to discern owing to the relative distance of the AP from the sinus node compared to that of the AV node and the relative conduction velocities of these competing AV conduction pathways. In such instances, adenosine usually prolongs AVN conduction time without affecting conduction time over the AP, thus making the preexcitation more evident electrocardiographically. **Figure 6.1** is such an example. Note that the QRS becomes more preexcited but the PR interval doesn't change, and this is related to prolonging AVN but not AP conduction time.

We note that some AP variants with decremental properties may be responsive to adenosine. In addition, an AP obtunded or incompletely ablated by radiofrequency ablation may be identified for further treatment by transient return of conduction after adenosine, presumably as a result of adenosine-induced hyperpolarization.

Dual AVN pathway physiology may be exposed in sinus rhythm with the administration of adenosine. The time course of adenosine effect on fast and slow pathways is usually different, with the fast AVN pathway frequently blocking earlier and for a longer time than the slow pathway. This may manifest as a sudden increase in the PR interval for a few beats, resulting from shift from fast to slow pathway prior to AV block. In the example in **Figure 6.2**, the PR interval increases suddenly after the sixth cycle with appearance of a retrograde P wave (arrow) during the expected adenosine effect. Adenosine caused block in the anterograde fast AVN pathway with shift to the slow pathway with a subsequent atrial echo going up the retrograde fast pathway. Subsequent block in the anterograde slow pathway after the echo prevented sustained AVNRT.

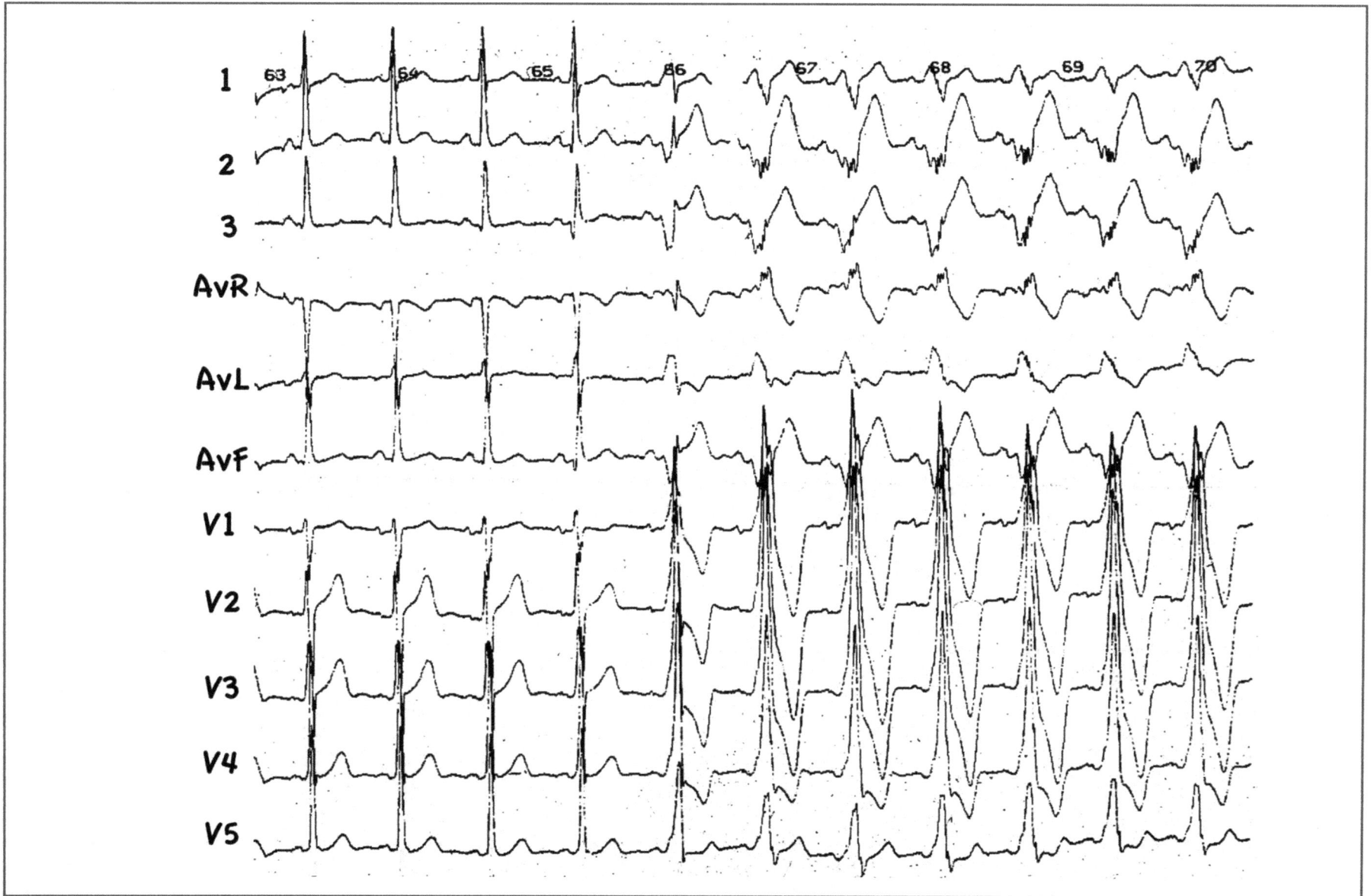

Figure 6.1

**Figure 6.2**

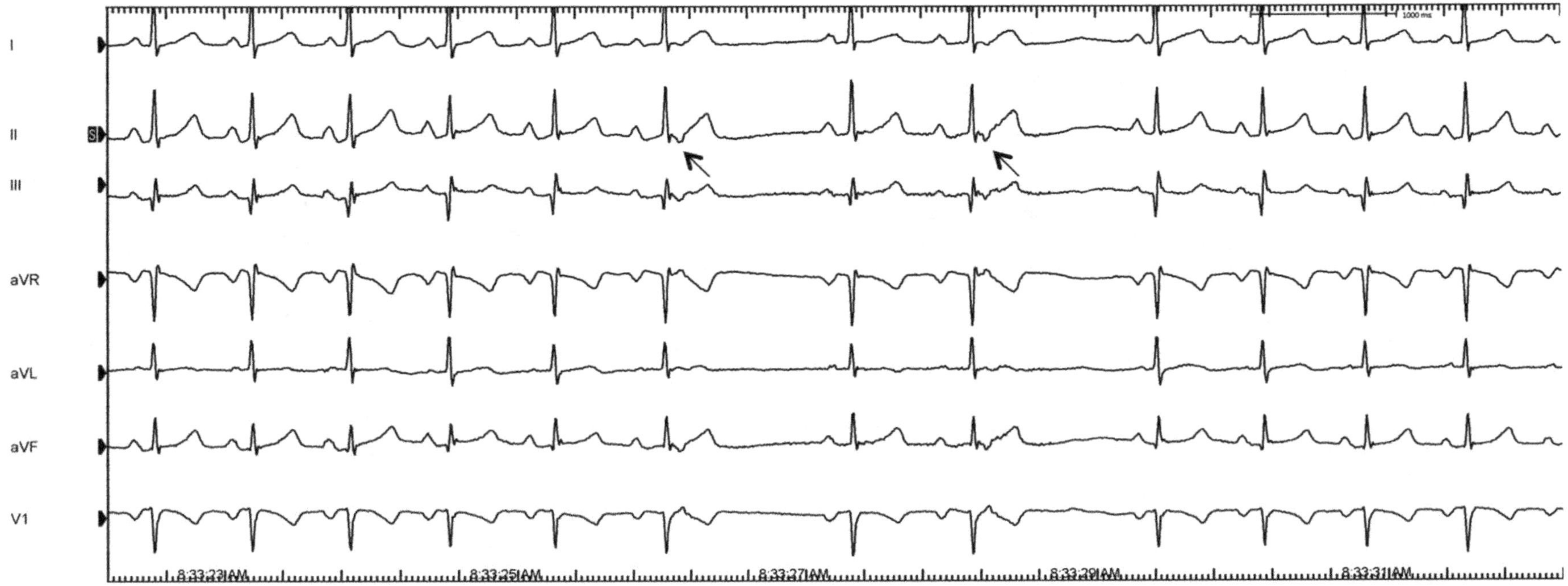

# Wide QRS Complex Tachycardia Diagnosis

Adenosine can be very helpful because most VT is not affected by adenosine. Sustained monomorphic VT is usually related to underlying scar, either ischemic or cardiomyopathic. Scar-mediated VT is not sensitive to adenosine and should continue unperturbed. On the other hand, SVT with aberrant conduction will declare itself by termination or rate slowing for an AVN dependent circuit or continuation of AT with transient AVN block for an AVN independent SVT-like atrial flutter. Although very useful as a test, a note of caution in that not all AV nodes will respond to adenosine, and conversely, some AT and VT will be adenosine sensitive. Idiopathic VT originating from the ventricular outflow region is believed to be due to triggered activity in many instances, occurs under increased sympathetic tone, and can be terminated by adenosine. In contrast, idiopathic fascicular VT from the LV septum is typically verapamil sensitive but not amenable to adenosine termination.

**Figure 6.3A** shows a wide-complex tachycardia with P waves in the diastolic interval with VT and SVT both plausible, that is, VT with retrograde conduction or the converse.

Adenosine induces temporary ventriculoatrial (VA) block (**Figure 6.3B**, asterisk) confirming the diagnosis of VT.

## Figure 6.3A

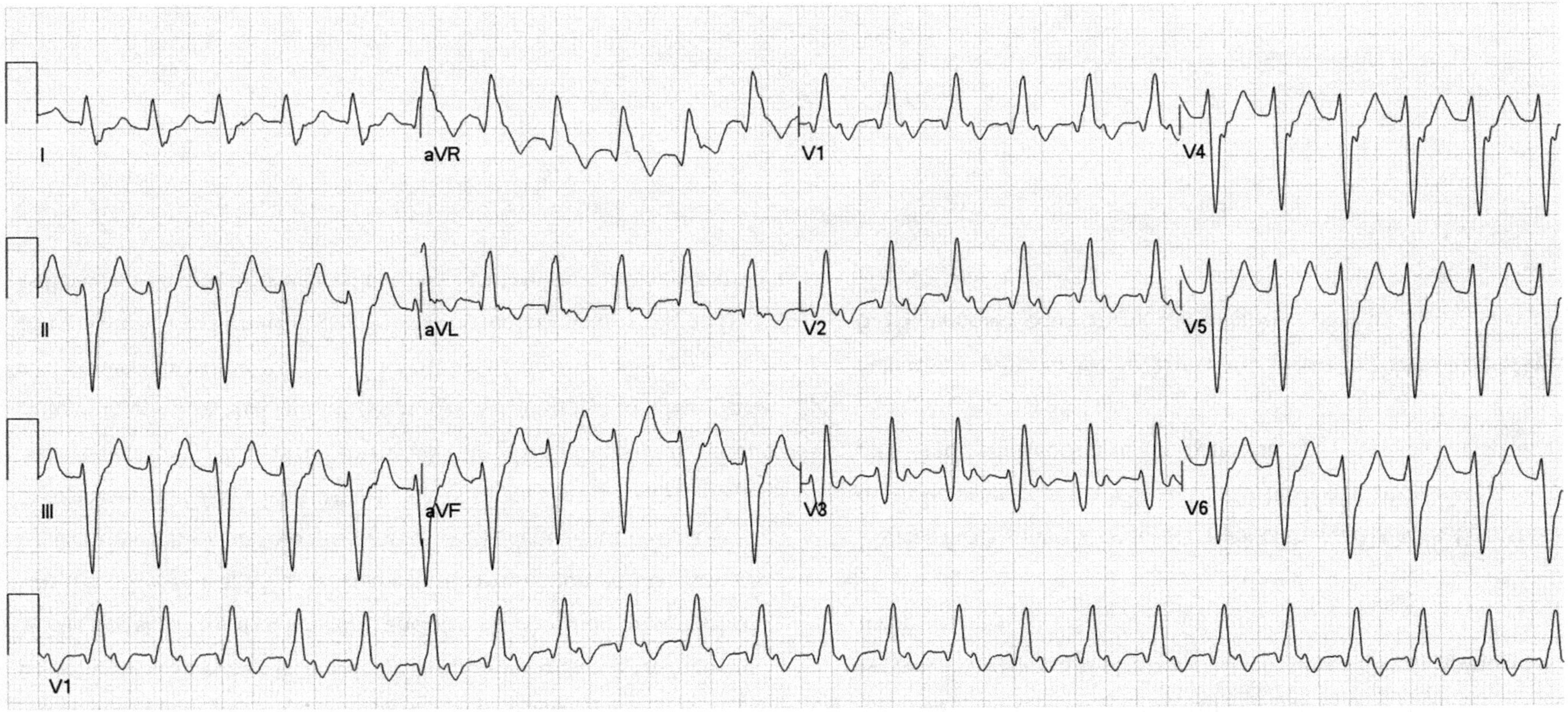

**Figure 6.3B**

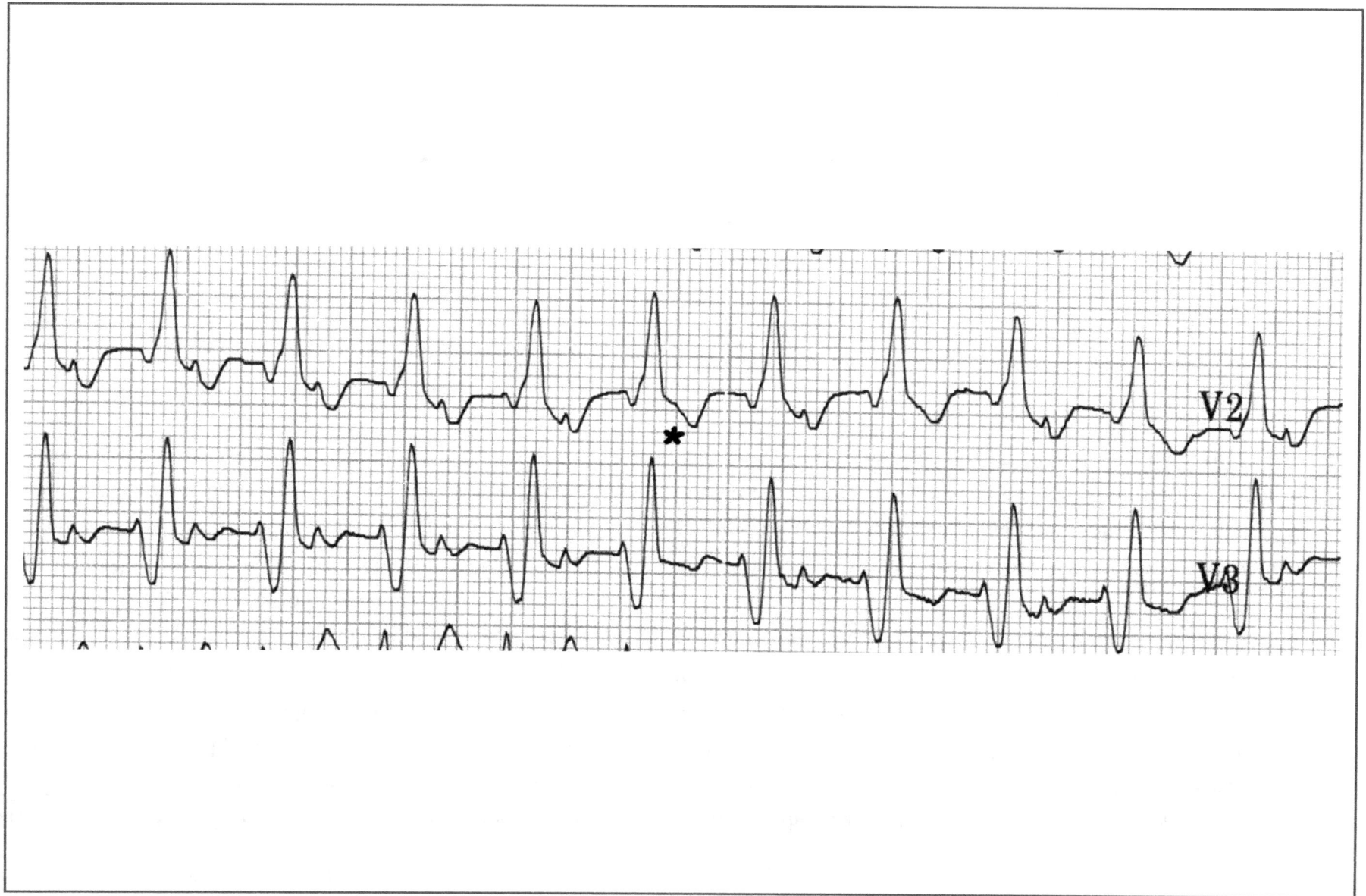

# SVT Responses to Adenosine

Typically, an AVN-dependent SVT will terminate or at least slow transiently (PR prolongation) with adenosine. There is no substantive slowing of the RR interval prior to break in the example in **Figure 6.4** suggesting AT, but the tracing has substantive artifact and the termination is obscured by ventricular ectopy. Termination may be followed by transient premature ventricular contractions (PVCs) and nonsustained VT as in Figure 6.4 and rarely even ventricular fibrillation (VF)! Ventricular irritability has been attributed to postadenosine surge in sympathetic tone. Atrial flutter and atrial fibrillation (AF) can also be seen after adenosine possibly related to consequent shortening of atrial refractoriness.

**Figure 6.5** shows an example of termination of an AT after adenosine. AV block occurs suddenly with minimal PR prolongation in the last conducted cycle before block and continues for 4 cycles before the AT itself terminates, and the last atrial beat seen is a sinus P wave. Thus, the adenosine effect on the AV node occurred first and the AT stopped abruptly without slowing of the AT rate.

The example in **Figure 6.6**, on the other hand, shows termination of an AT with slowing of the atrial rate and minimal effect on the PR interval. (We note that the P-wave orientation in this example is "high to low" with positive P vectors in leads II and III, ruling out VA conduction during an AVN dependent SVT.)

**Figure 6.4**

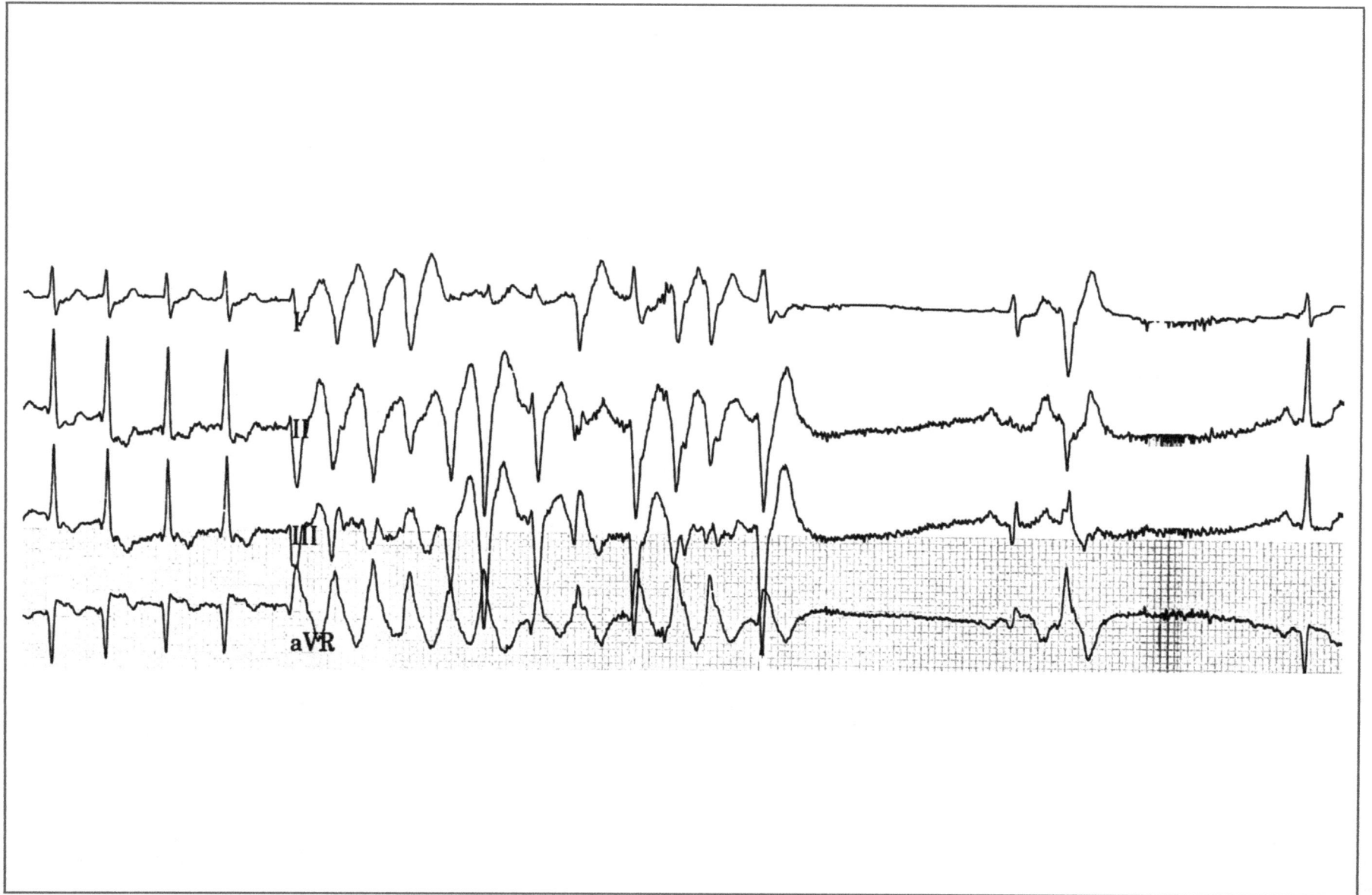

**Figure 6.5**

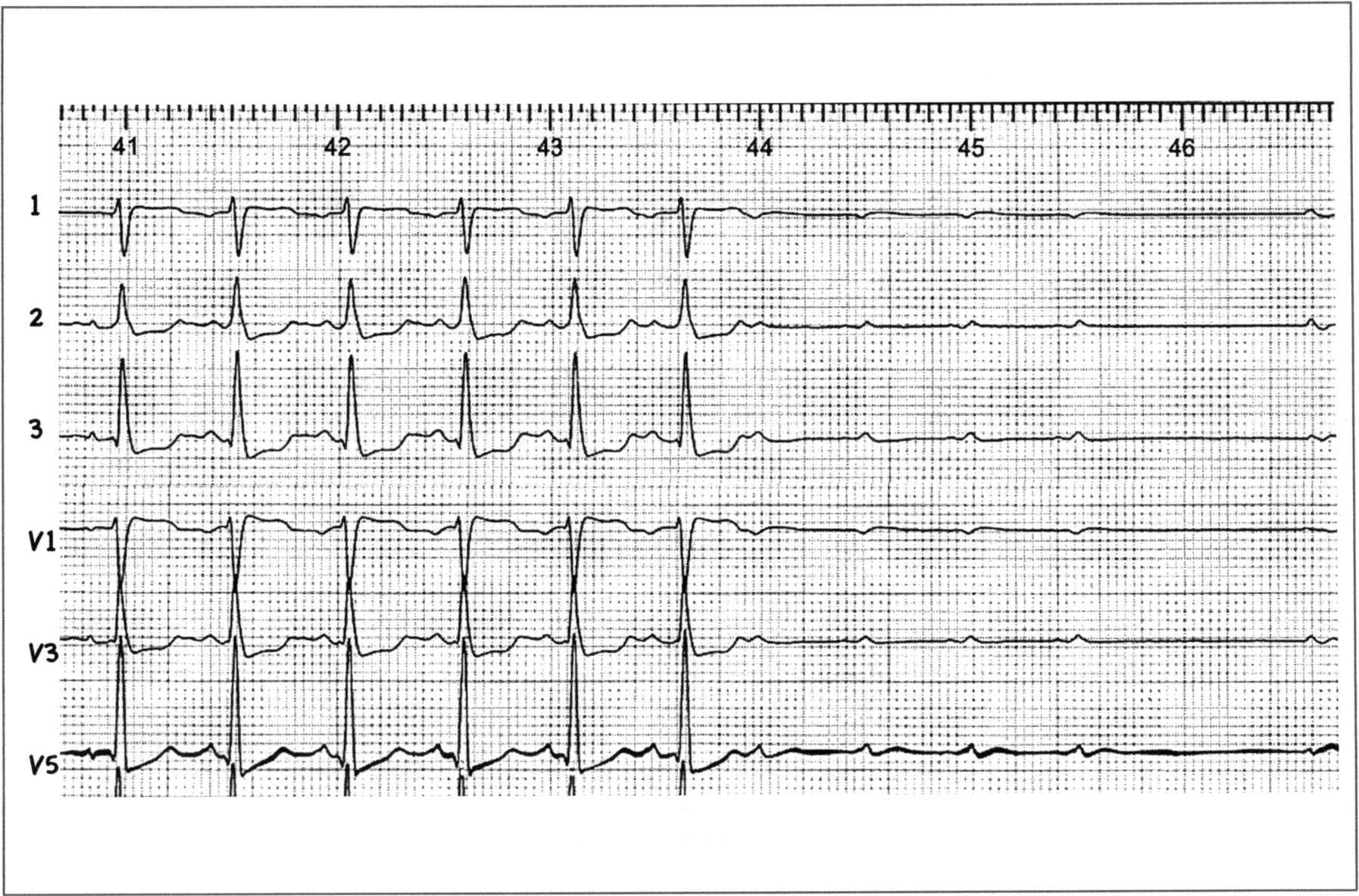

**Figure 6.6**

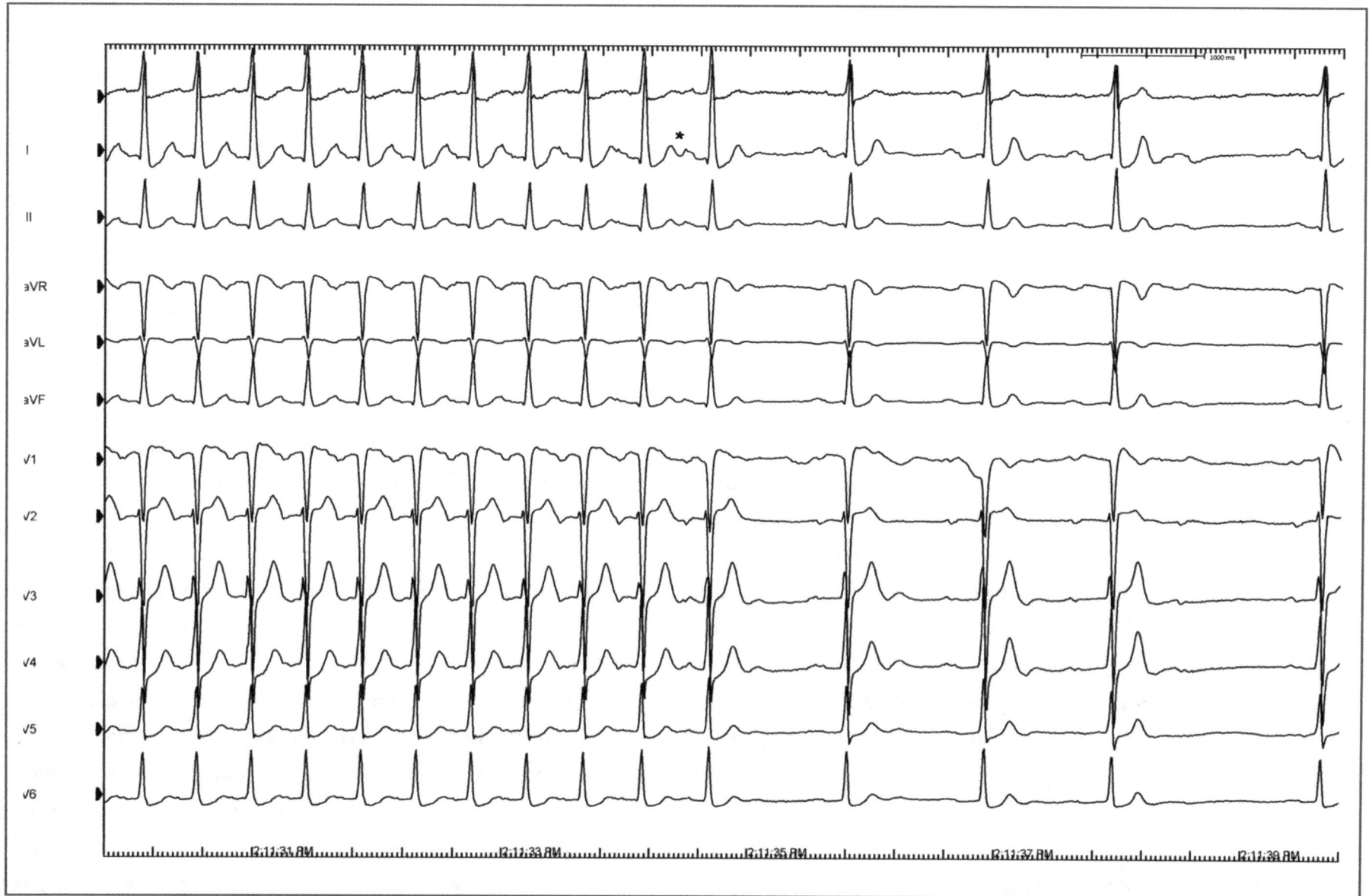

**Figure 6.7** shows another SVT after adenosine. The atrial cycle length (CL) prolongs during the top strip with no substantive change in the PR interval and this is evident even without careful measurement. This is followed by sinus rhythm with AV block and a dissociated junctional escape as seen in the second strip. There is early recurrence of AT (third strip), which starts with a late-coupled PAC of the same P morphology as the AT. This was AT but, strictly speaking, one could not rule out a long RP tachycardia with retrograde conduction over an adenosine sensitive structure (either AVN or an atypical AP).

*Thus, the differential time course between the AVN effect of adenosine and its effect on an adenosine-sensitive AT usually assists in differentiating an AV node-dependent SVT from an adenosine-sensitive AT.*

SVT in **Figure 6.8A** is a "long RP" tachycardia that could be AT, atypical AVNRT, or AVRT with a decremental AP as the retrograde limb of AVRT. The pattern of termination (**Figure 6.8B**) with adenosine virtually rules out AT. The tachycardia terminates with a P wave (asterisk). This is possible with AT *only if* AV node block occurs at the same instant. The tachycardia slows with prolongation of the VV interval *preceding* the prolongation of the AA interval. Note that the PR interval prolongs on the last QRS before the break and the delay in arrival of the last A prior to the break only occurs *after* this (i.e., VV change precedes AA change).

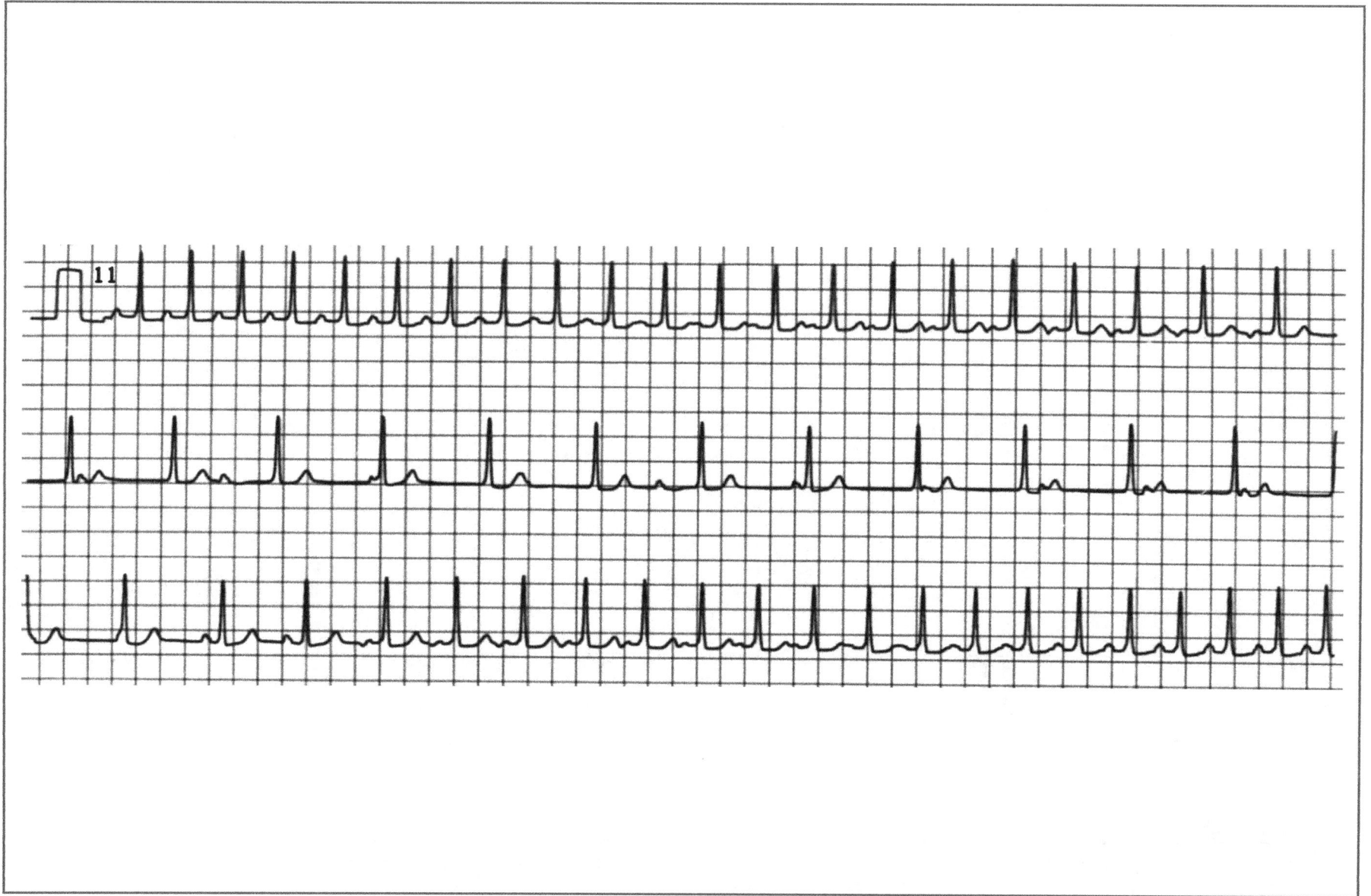

Figure 6.7

**Figure 6.8A**

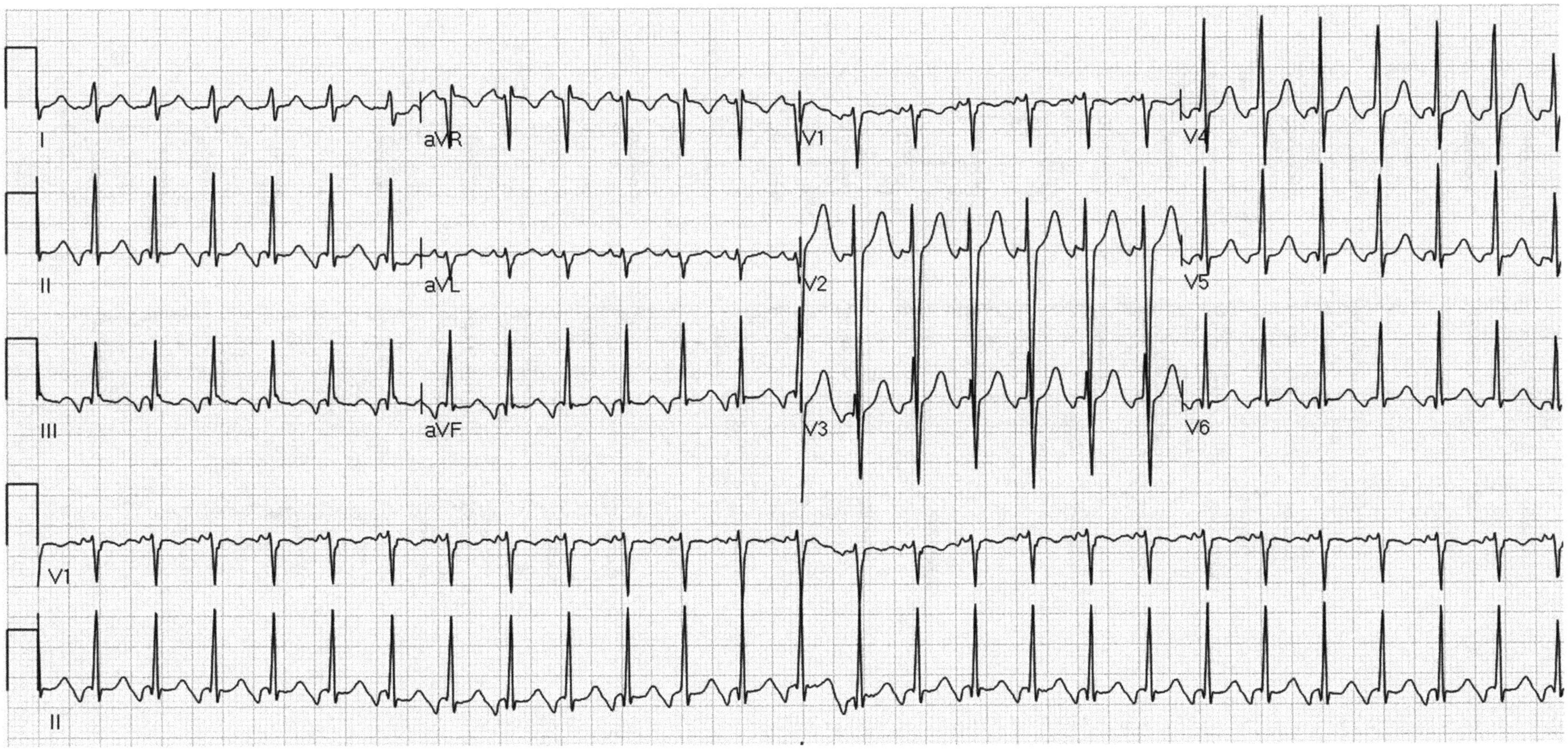

**Figure 6.8B**

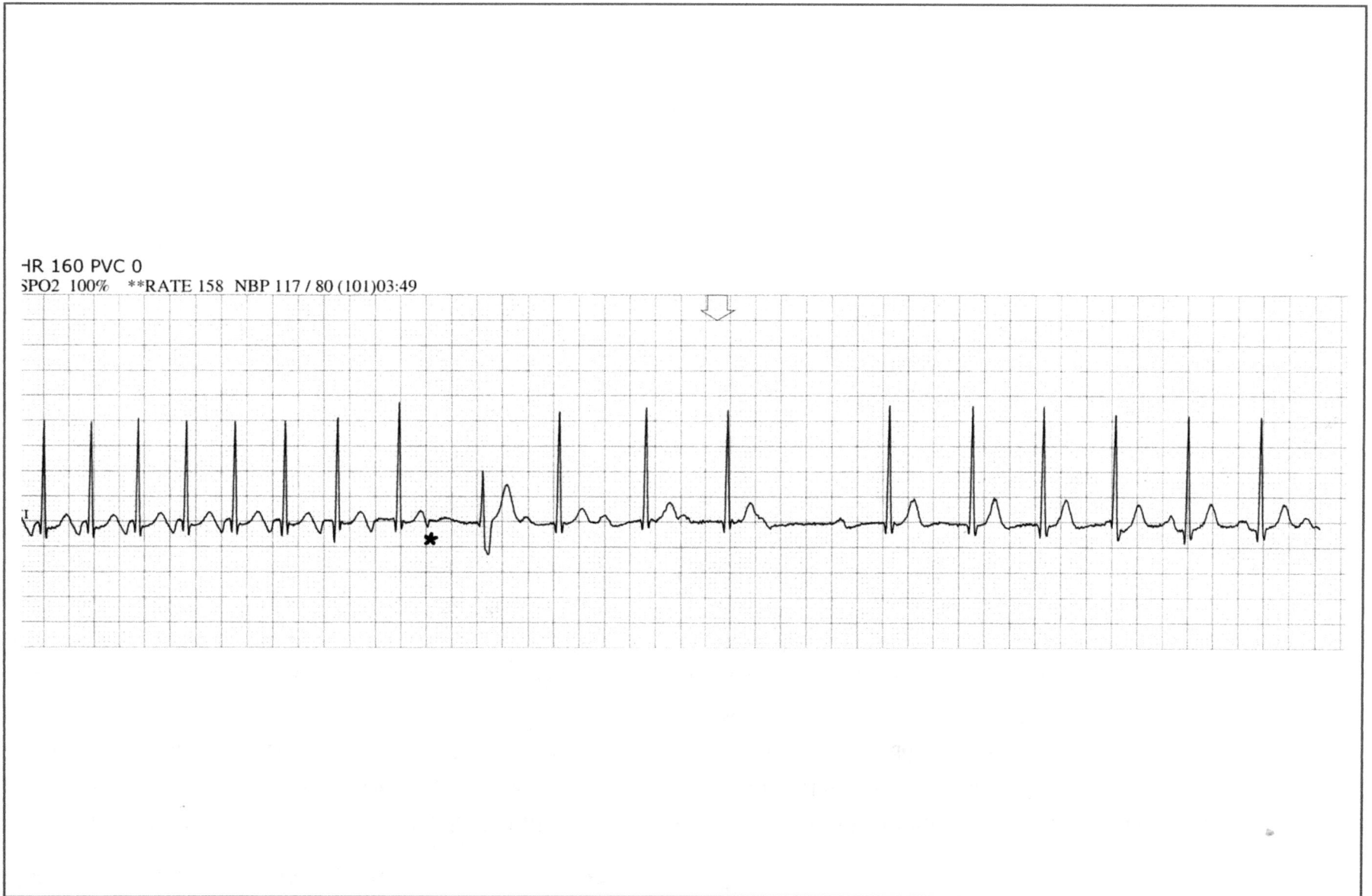

HR 160 PVC 0
SPO2 100%  **RATE 158  NBP 117 / 80 (101)03:49

# Differentiating AVNRT from JT

Distinguishing AVNRT from junctional tachycardia (JT) with pacing maneuvers is generally not problematic but adenosine response may be useful when there are no intracardiac leads available. **Figure 6.9** illustrates such an example with early retrograde atrial activation potentially resulting from one of the following: (1) AT with a long PR interval, (2) AVNRT, or (3) JT with retrograde conduction. *Although the intracardiac electrograms are retained for clarity, they are not necessary and atrial activity is very clear from the surface ECG (arrow head).* Adenosine results in VA block (asterisk) with ongoing JT, which slows modestly before terminating. This clearly rules out AT. (Theoretically, one can consider that AVNRT does not require atrial activation and adenosine can cause block from the AVN to the atrium without immediately affecting AVNRT. This would be a bit of a stretch here and this indeed was JT.) This type of response in JT is not unusual in our experience, although never formally evaluated to our knowledge.

**Figure 6.9**

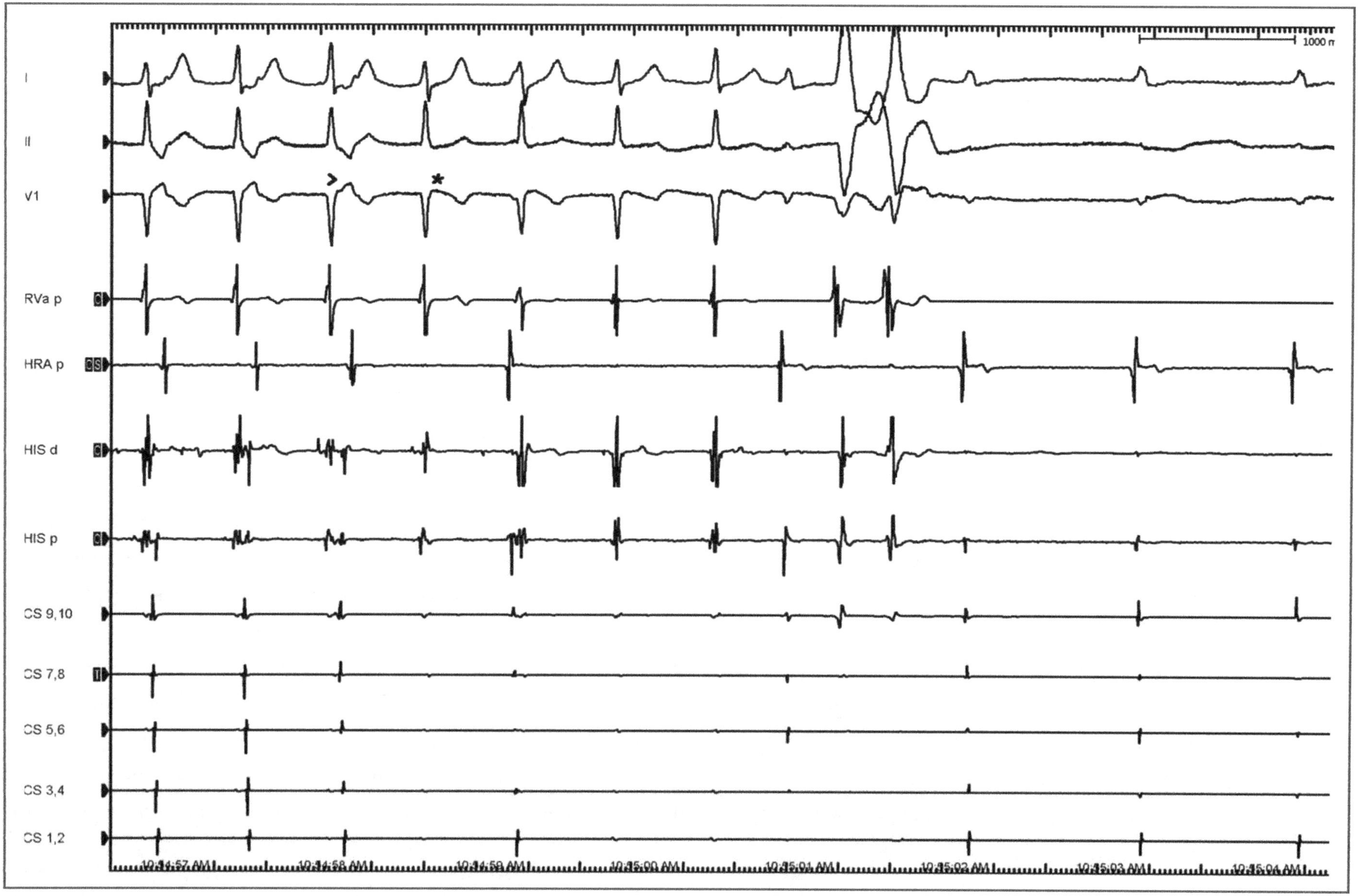

# Adenosine and Recovery of Conduction

Recurrence of conduction in APs or atrial tissue after apparently successful ablation has been thought to be related in some cases to incomplete ablation or obtunding affected tissue only, allowing recovery. Restoration of conduction in the affected tissue transiently (**Figure 6.10**, compliments of Dr. Simon Modi) has been suggested to indicate that the tissue is still viable and more ablation is required. In our example, a circular mapping electrode is seated in the left upper pulmonary vein (PV). The surface leads initially show sinus rhythm *outside* of the PV and AF *within* the PV. Adenosine transiently restores conduction from the PV to the atrium, resulting in transient overt AF. Further ablation was done until adenosine could no longer do this. The mechanism of transient restoration of conduction is thought to be hyperpolarization of the transmembrane potential.

The clinical significance of demonstrating adenosine-induced recovery of conduction is still not rigidly proved but many physicians target these tissues for further energy application.

Figure 6.10

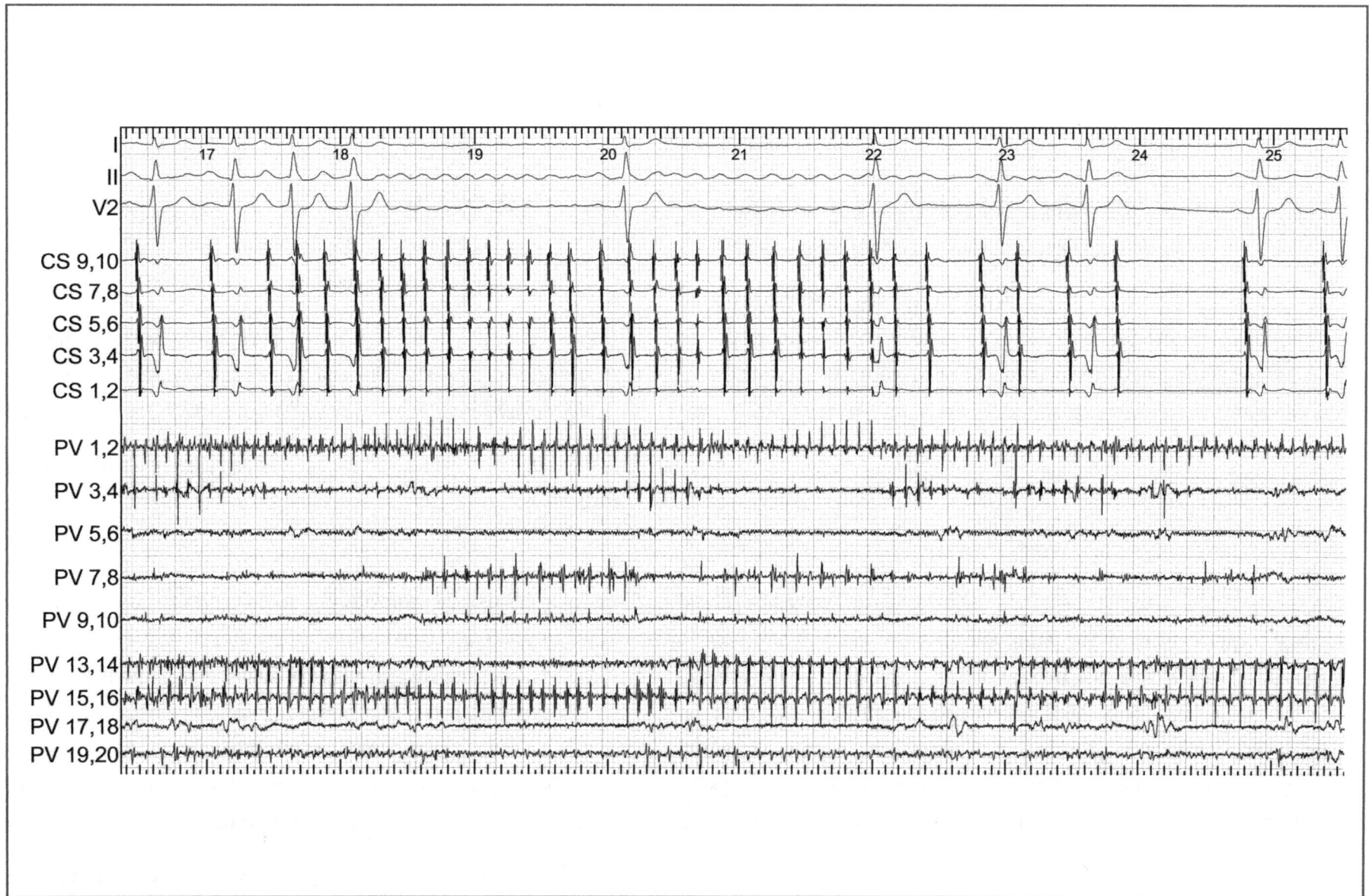

## Miscellaneous Nonpacing Maneuvers

Other currently available pharmaceuticals are occasionally useful in the EP laboratory. Verapamil and diltiazem, among their effects, prolong AVN conduction time and refractoriness for a more prolonged time than adenosine. The addition of beta-blockers to verapamil substantively augments this effect. This has been occasionally helpful to us in blocking retrograde AVN conduction to assist in ablation of septal APs that had very similar retrograde atrial activation sequence.

A potentially useful tool diagnostically is the cryoablation catheter. This has the ability to cool tissue adjacent to it and stop conduction in tissues cooled to less than 0°C. For example, one could evaluate the functional significance of a mapping site during VT where slowing or cessation of VT would validate the functional importance of the tissue underneath to the abnormal circuit. This could be a very powerful maneuver indeed!

## Suggested Readings

1. Josephson ME, Seides SF. *Clinical Cardiac Electrophysiology: Techniques and Interpretations*. Philadelphia, PA: Lippincott Williams & Wilkins; 2008.

2. Lerman BB, Belardinelli L. Cardiac electrophysiology of adenosine. Basic and clinical concepts. *Circulation*. 1991;83:1499–1509.

3. Iwai S, Markowitz SM, Stein K, et al. Response to adenosine differentiates focal from macroreentrant atrial tachycardia: validation using three-dimensional electroanatomic mapping. *Circulation*. 2002;106:2793–2799.

4. Mallet ML. Proarrhythmic effects of adenosine: review of the literature. *Emerg Med J*. 2004;21:408–410.

# Index